Behind The Clouds

The stars that shine

Shruti Riya

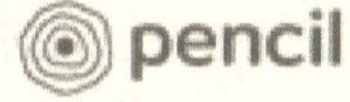

ISBN 978-93-5458-516-6

Published in India 2021 by Pencil

A brand of
One Point Six Technologies Pvt. Ltd.
123, Building J2, Shram Seva Premises,
Wadala Truck Terminal, Wadala (E)
Mumbai 400037, Maharashtra, INDIA
E connect@thepencilapp.com
W www.thepencilapp.com

Author biography

Shruti Riya is a high school from Jamshedpur city. She is a passionate star grazer and has deep interest of astronomy.

CONTENTS

Epigraph

The stars & the constellations are the shelters of glitter which provide a home to the homeless.

It's true love, not lust;
An Opinion, not a judgment;
The genuine care, not brag;
The late-night talks of the loved one remained untold.

The emotion and feeling of being bound with anyone, similar to you are the grateful feelings. The shining glitters contain a message for us, the message of being alive... When someone dear to us criticizes us, we shatter into a million pieces. The stars resemble those shattered pieces, And the constellations join them. The indication of not being alone. The night sky is the place that takes away loneliness and provides an indication of belonging to a bigger; world with many people like us.

They share their story & We share ours.

Foreword

There are 48 constellations discovered by Ptolemy. Those 48 constellations are associated with mythologies. There are various myths for all the constellations, i.e., Greek mythology, Egyptian mythology, Eastern mythology, and many more. Sometimes the story in one myth is different from that in another. Also, sometimes two constellations have the same or similar story.

History was written on a piece of paper by our ancestors. Lack of access to those papers sometimes causes misconceptions about the past.
The future is also unpredictable. We all are in motion. Our planet, our solar system, the galaxy, everything is in constant motion. Things are dynamic. Years will pass, and one day, this location of stars will become history.

The author had tried her best to put all the scientific and mythological details. The myth might not be the same as those in certain books, but it's one, which was told by great historians. The scientific location might change after years but it is true now.

Preface

Constellation
A constellation is a group of stars that forms a pattern and has a name.

There are 8 families in the constellation.

The Ursa Major Family of the constellation- Ursa Major, Ursa Minor, Draco, Bootes, Camelopardalis, Canes Venatici, Coma Berenice, Corona Borealis, Lynx, and Leo Minor.

The Zodiac Family of the constellation- Aries, Piscis, Taurus, Gemini, Cancer, Leo, Virgo, Libra, Scorpius, Sagittarius, Capricornus, and Aquarius.

The Perseus Family of the constellation- Perseus, Andromeda, Cassiopeia, Cepheus, Cetus, and Pegasus.

The Hercules Family of the constellation- Hercules, Aquila, Lyra, Sagitta, Cygnus, Vulpecula, Sextans, Hydra, Corvus, Crater, Serpens, Ophiuchus, Centaurus, Scutum, Ara, Corona Australis, Lupus, Triangulum Australe, and Crux.

The Orion Family of the constellation- Orion, Canis

Major, Canis Minor, Lepus, Monoceros.

The Heavenly waters Family of the constellation- Delphinus, Equuleus, Eridanus, Piscis Austrinus, Columba, Pyxis, Carina, Puppis, Vela.

The Johann Bayer Family of the constellation- Apus, Hydrus, Dorado, Volans, Pavo, Grus, Phoenix, Indus, Tucana, Musca, and Chamaeleon.

The Lacaille Family of the constellation- Norma, Circinus, Telescopium, Fornax, Sculptor, Microscopium, Horologium, Reticulum, Caelum, Mensa, Octans, Pictor, Antlia.

Acknowledgements

First of all, I would like to thanks my parents for their love, patience, and support. Dear Mother and Father, you always supported me and follow my dreams. I will be thankful to you for my entire life.

Secondly, I would like to thanks my brother and sister, they both provided me with warmth and support. The support to do achieve everything I desired, to work for all and everything, I ever dreamt of. I have no words to express my gratitude towards them. I am forever indebted to you.

I would like to thank the stars, especially Sirius and Arcturus. The twinkling star never left me alone. In the time of failure, they provided me hope.

Andromeda

Andromeda, the chained woman, is the 19th largest constellation located in the first quadrant of the Northern Hemisphere (NQ1). It occupies an area of 722sq. Degree in the sky. It was listed by the 2nd BC Greek astronomer Claudius Ptolemy.

You might hear the story of beauty and beast but have you ever heard of a title named, Beauty and her mother? Of course not, because no such title ever exists. But a leading gorgeous girl is much related to this.

Andromeda, daughter of Cepheus and Cassiopeia, the young princess of Ethiopia. Her overwhelming beauty was praised by everyone. Her curls were enough to trap the emotional heart of any prince of the world. Andromeda was not much impressed by her look. Her happiness lies in saving others. She always seemed like a warrior rather than a princess decorating herself. People saw her green maleficent eye, but she saw the red flame burning inside her. Being a girl, she was not allowed in aggressive war but her stubbornness brought her into it. She used to dress as a man to enters into war. Cassiopeia was very much galvanized by her daughter's beauty and had a habit of instilling others. This habit turned into a curse for the Ethiopians. Cassiopeia insulted sea nymphs (female spirits of the seawater), saying that her daughter was exceedingly prettier than them; and inflamed them to start the dismantling of Ethiopia. The passionate sea nymphs complained this to Poseidon, the god of the sea. Poseidon ordered Cetus to destroy Ethiopia. Cetus, the fear of everyone, was the sea monster. His broad terrifying teeth were scary enough to frighten anyone.

To extricate his kingdom from the sea monster, King Cepheus sought the Oracle of Ammon. The Oracle of Ammon advised the king that the only way to save his kingdom is to mollify the sea nymphs and the sea monster. He also told him to sacrifice Andromeda to them. After that, Andromeda was chained to a rock near the sea as a sacrifice to the sea monster. At the time of sacrifice, Perseus, the son of Zeus and Danae, was moving towards Ethiopia. He had killed Medusa and possessed her head with him. On his way, he found Andromeda chained to the rock. Someone narrated the entire story. Perseus went to the castle and asked Cepheus that if he kills the sea monster, then he could marry Andromeda. Cepheus happily agreed. Perseus went to save Andromeda. He grasped the medusa's head and turned the Cetus into stone. Then, Andromeda and Perseus lived happily.

It belongs to the Lacaille family of the constellation. The neighboring constellations are Cassiopeia, Lacerta, Pegasus, Perseus, Pisces, and Triangulum. The major stars are: Alpheratz (Alpha Andromedae), Mirach (Beta Andromedae), Almach (Gamma Andromedae), Adhil (Xi Andromedae), Nimbus (51 Andromedae); some notable deep sky objects are - Andromeda Galaxy (NGC 224);

Andromeda satellites; Andromeda's Cluster (Mayall II); Blue Snowball Nebula (7662); Ghost of Mirach (NGC 404). It is associated with the Andromedids meteor shower in November. The best time to view the constellation at 9:00 pm in November between +90 degrees and -40 degrees.

Antlia

Antlia, the pump, is the small faint constellation in the southern sky. It is the 62nd largest constellation in the sky. It covers an area of 239sq. degree in the second quadrant of the southern hemisphere (SQ2). It was created by the French astronomer Nicolas Louis de Lacaille.

The French Astronomer Nicolas-Louis de Lacaille while on an exploratory trip to the Cape of Good Hope created this constellation. He named the constellation to commemorate the remarkable discovery of an air pump by the French physicist Danis Papin. Its original name is Antlia Pneumatica and in French is Machine Pneumatique. It represents the single-cylinder pump that Papin used in his practical experiments in 1670. Its name is connected to technological discovery and symbolizes the Age of Enlightenment.

It belongs to the Lacaille family of the constellation. It is bordered by Hydra, Centaurus, Pyxis, and Vela. The notable stars in the constellation are Alpha Antliae, Epsilon Antliae, Lota Antliae, Theta Antlia, and Eta Antlia. The major deep sky objects are the Antlia Dwarf; NGC 2997 (ESO 434-G 35) an unbarred spiral galaxy; Antlia Cluster (Abell S0636); IC 2560 a spiral galaxy; IC 2545 a pair of merging galaxy; NGC 3125 a starburst galaxy, and

NGC 3244 a spiral galaxy. The constellation is best visible from January to March and can be only viewed from the southern hemisphere. This constellation is not associated with any meteor shower.

Apus

Antlia, the pump, is the small faint constellation in the southern sky. It is the 62nd largest constellation in the sky. It covers an area of 239sq. degree in the second quadrant of the southern hemisphere (SQ2). It was created by the French astronomer Nicolas Louis de Lacaille.

The French Astronomer Nicolas-Louis de Lacaille while on an exploratory trip to the Cape of Good Hope created this constellation. He named the constellation to commemorate the remarkable discovery of an air pump by the French physicist Danis Papin. Its original name is Antlia Pneumatica and in French is Machine Pneumatique. It represents the single-cylinder pump that Papin used in his practical experiments in 1670. Its name is connected to technological discovery and symbolizes the Age of Enlightenment.

It belongs to the Lacaille family of the constellation. It is bordered by Hydra, Centaurus, Pyxis, and Vela. The notable stars in the constellation are Alpha Antliae, Epsilon Antliae, Lota Antliae, Theta Antlia, and Eta Antlia. The major deep sky objects are the Antlia Dwarf; NGC 2997 (ESO 434-G 35) an unbarred spiral galaxy; Antlia Cluster (Abell S0636); IC 2560 a spiral galaxy; IC 2545 a pair of merging galaxy; NGC 3125 a starburst galaxy, and

NGC 3244 a spiral galaxy. The constellation is best visible from January to March and can be only viewed from the southern hemisphere. This constellation is not associated with any meteor shower.

Aquarius

Aquarius, the cupbearer, is the 10th largest constellation in the sky. It covers an area of 980sq. Degree in the fourth quadrant of the southern hemisphere(SQ4). It was listed in the 48th Ptolemy's constellation in the 2nd century BC. It represents a single vase from which the stream is poured down to the Pisces Austrinus and the region where it lies is referred to as the sea as it contains the constellation associated with water: Pisces- the fish; Eridanus- the river; Cetus- the whale.
In Greek mythology, Ganymede was a divine hero, whose homeland was Troy. He was the son of the king of Tros and Callirrhoe. A young Trojan boy with an elegant look and was the most beautiful among the mortals. Zeus was attracted by him and desired him to be a cupbearer of God.

Ganymede was kidnapped by Eos, the goddess of Dawn. The goddess' affection for the young boy led her to kidnap the boy. Zeus, who was already impressed and interested in Ganymede, ordered his Eagle Aquila to steal the boy from Eos and bring him to Mount Olympus. Zeus saved the boy from the goddess and employed him as 'the water bearer'. From then, his work was to bear water and serve gods. In a short time, he was tired of his work. He then refused Zeus to serve the Gods. Enduring kind to him,

Zeus placed Ganymede among the stars as Aquarius.

It belongs to the Zodiac family of the constellation. The neighbouring constellation is Aquila, Pisces, Pegasus, Equuleus, Delphinus, Cetus, Sculptor, Capricornus, Pisces Austrinus. The major named stars of the constellation are- Sadalsuud (Beta Aquarii); Sadalmelik (Alpha Aquarii); Skat (Delta Aquarii); Sadachbia (Gamma Aquarii); Sadaltager (Zeta Aquarii); R Aquarii. The other notable deep sky objects are- Messier 2 (NGC 7089) and Messier 72 (NGC 6981) are the globular clusters; Messier 73 (NGC 6994) an asterism of stars; Saturn Nebula (Caldwell 55) and Helix Nebula (Caldwell 63) are planetary nebulae; the Aquarius Dwarf (DDO 210) and Atoms for Peace Galaxy (NGC 7252) are the Dwarf Galaxy. It is also associated with four meteor showers: March Aquariids, Delta Aquariids, Eta Aquariids, and Iota Aquariids. It is best visible in October between +65 degrees and -90 degree latitudes in the sky.

Aquila

Aquila, the eagle, is the 22nd largest constellation in the sky located in the fourth quadrant of the Northern hemisphere (NQ4). It covers an area of 652sq. degrees in the sky. It was cataloged by the Greek astronomer Claudius Ptolemy in the 2nd century BC.

It represents the eagle of the thunder god Zeus. According to classic Greek, Aquila was the eagle that faithfully carried the thunderbolts of Zeus. Zeus observes a Trojan boy, whom he desired to become a wine pourer of the Gods. Zeus forced the eagle to retrieve Ganymede, a son of one of the kings of Troy to Mount Olympus to serve as cup-bearer to the gods.

It belongs to the Hercules Family of the constellation. The neighboring constellations are Aquarius, Capricornus, Sagitta, Hercules, Ophiuchus, Serpens Cauda, Sagittarius, and Delphinus. The notable stars of the constellations are Altair (Alpha Aquilae), Alshain (Beta Aquilae), Tarazad (Gamma Aquilae), Deneb el Okab (Epsilon Aquilae), Bezek (Eta Aquilae). The notable deep sky objects in the constellations are Phantom Streak Nebula (NGC 6741); Barnard 142 and 143 (E Nebula); The Glowing Eye Nebula (NGC 6751). It is associated with two meteor showers- June Aquilids and Epsilon Aquilids. It is best visible in August between +90 degree and -75 degree latitudes.

Ara

Ara, the altar, is the 63rd largest constellation as it covers an area of 237sq. Degrees in the sky. It is located in the third quadrant of the southern Quadrant (SQ3). It was created by the 2nd century Greek Astronomer Claudius Ptolemy.

Ara signifies the altar on which Zeus and the others vowed to Titans and overthrow Cronus, the ruler of the Universe. Cronus dethroned his father Uranus. A Prophecy predicted that the future and said that the same fate would befall Cronus. He would also be defeated by his son. The prediction exasperated Cronus, and to prevent it from happening, he decided to kill all his children. Soon, he swallowed all of his children. Firstly Hestia, then Demeter, Hera, Hades, and Poseidon. Then, it was the turn of the youngest, Zeus. When Zeus was born, his mother Rhea, hiding him from everyone, sent him to Crete. Rhea arranged a stone for Cronus, mentioned to him the stone was his son. Zeus was miraculously saved.

When Zeus grew up, he made Cronus emesis his brothers and sister. Then Zeus and his beloved brother & dear sister swore to overthrow Cronus and the Titans. The fierce war between the Gods and Titans was engaging and had typically lasted for a decade. The Gods won in the end. Zeus became God of the sky, Poseidon the ruler of the Sea, and Hades the king of the underworld. Zeus placed the altar among the stars to commemorate God's victory.

This constellation is bordered by Apus, Corona Australis, Norma, Scorpius, Telescopium, Triangulum Australe. The major stars are- Beta Arae, Alpha Arae, Gamma Arae, Zeta Arae, Mu Arae, Epsilon Arae. The remarkable deep-sky bodies in the constellation are the Stingray Nebula (Hen 3-1357); Ara Cluster (Westerlund 1); Water Lily Nebula. This constellation is not associated with any meteor shower. The best time to view the constellation is the month of July between +25 degrees and -90 degrees latitude.

Aries

Aries, the ram, is the 39th largest constellation occupying an area of 441 sq. degrees in the sky. It is located in the first quadrant of the Northern Hemisphere (NQ1). It was cataloged by the 2nd century BC Greek astronomer Claudius Ptolemy. It represents a ram's horns.

The constellation is associated with the golden ram that rescued Phrixus and Helle. The golden ram was the blood of Poseidon.

King Nephele was married to Athamas and gave birth to two children, a boy Phrixus and a girl, Helle. In a couple of shakes, they got divorced, and the second wife of Nephele was thirsty for the blood for his children. Nephele' second wife had conspired against his stepson. Nephele, realizing that his children are in danger, was worried. Hermes sends the golden ram to rescue them, the youngster sat on it and they flew from their wicked stepmother. After the rescue, They sacrificed the golden ram in the incredible honour of the mighty gods to Zeus. Zeus placed the ram in the sky for is most famous for his help.

It belongs to the Zodiac Family of the constellation. Its neighbouring constellations are- Perseus, Triangulum, Pisces, Cetus, and Taurus. The major stars are- Hamal

(Alpha Arietis), Sheratan (Beta Arietis), Mesarthim (Gamma Arietis), Botein (Delta Arietis), and Bharani (c Arietis). It is associated with six meteor showers- Aries Triangulids, Autumn Arietids, Delta Arietids, Daytime-Arietids, Epsilon Arietids, and May Arietids. It is best visible in December.

Auriga

Auriga, the charioteer, is the 21st largest constellation as it covers an area of 657 degrees in the sky. It is located in the first quadrant of the Northern hemisphere (NQ2). It was described by the 2nd-century Greek astronomer Claudius Ptolemy and listed in his 48 constellations.

In Greek mythology, Hephaestus was the god of fire, metalworking, stone masonry, forges, and art of sculpture and the son of the goddess Hera and Zeus. He was created by his mother only and was born deformed. He was a terrible mortal among perfectly beautiful immortal gods. He was forced to leave Mount Olympus as he was not liked by his own parents. He possessed some of his mother's power. He left Olympus and set up his workshop under a volcano. He produced the most superior weapons in Greece.

Later, Zeus accepted and regarded him as the god of fires and as a reward, married him to the Aphrodite, the god of love, beauty, pleasure, passion, and procreation, to prevent the war of fighting hands. Hephaestus was unsatisfied with the reward. His wife Aphrodite, a young attractive woman accompanied by the winged godling Eros, was in love with the god of war, Ares. Before long she became pregnant by Ares; her marriage with Hephaestus ended. She gave birth to five children- Eros, Anteros, Deimos, Phobos, and Harmonia. Hephaestus was badly cheated by the goddess of love.

During the Trojan Wars, it was Hephaestus's duty to produce the weapons for the wars. The goddess of wisdom, Athena frequently visited his workshop which makes him a little happier. Poseidon, the god of the sea and the most generous enemy of Athena, got a chance to avenge Athena. Poseiden lies to Hephaestus that Athena loves; on that night when Athena came to his workshop, he tried to seduce her but Athena already forewarned him. In that intercourse, Hephaestus' semen falls on Athena's thigh. She washed and it fell to the earth. Gaia accepted it and gave birth to a child named Ericthonius, his body was half man and half serpent. Athena adopted him as his child and raised him.

He was not like an ordinary legend, acquired many skills, and came to be the first man to tame and harness four horses to a chariot, following the chariot to the sun god. Zeus placed him in the sky for his bravery and creation as the constellation of Auriga.

It belongs to the Perseus Family of the constellation. The neighboring constellations are Perseus, Camelopardalis, Gemini, Lynx, and Taurus. The major stars are- Carpella (Alpha Aurigae) is the 6th brightest star in the sky; Menkalinan (Beta Aurigae); Mahasim (Theta Aurigae); Hassaleh (iota Aurigae. The notable deep sky objects are- the Flaming star Nebula (IC 405). It is associated with two meteor showers- Aurigids (September) and Delta Aurigids

(October). It is best visible between February and March between +90 degree to -40 degree latitude.

Boötes

Boötes, the herdsman, is the 13th largest constellation in the sky as it covers an area of 907sq. Degree in the sky. It was listed by the 2nd-century Greek astronomer Claudius Ptolemy.

Boötes mythology is associated with the story of Kallisto, the great bear in Greek Mythology. The myth of Boötes overlaps with the mythology of Ursa Major. The actual reason for not mentioning Greek mythology is the similarity. In Egyptian Mythology, it is associated with agriculture.

Callisto, a beautiful princess, and daughter of King of Arcadia was best known for her elegance. She attracted both men and Gods. The elegant princess became the lover of Zeus, and a boy Arcus was born from their union.

But miseries and distress were not far. After learning of another betrayal of Zeus, her wife Goddess Hera decided to punish Callisto. Hera cursed the princess and then fur started to appear on the young princess's soft skin. Her delicate hand turned into paws and sharp claws. The princess tried to screen for help but only a dreadful howl came out of her mouth. Callisto had been transformed into the great bear. The bear wandered through the forest and stood on her two paws and begged Zeus to have her original shape back. Zeus did not want to antagonize his wife and did nothing for Callisto.

The bear was now roaming around the bank of the forest, next to where she lived. She used to hide from everyone and watched her friends and relatives. Despite the fear of being snipped by the hunter, she only risked her life to see her son Arcus grow. The little boy grew up believing that one day he would see his mother again. Time passed and one day while wandering through the forest, she encountered a hunter who was her beloved son. The mother forgot that she was a bear and stood up and tried to hug her son. Arcus thought that the bear was about to attack him and thrust the spear into the bear. But Zeus did not allow such a tragedy to occur. God transformed both mother and son into a constellation. The mother became known by the name of Boötes.
But the Goddess Hera was furious to see her rival and her son being honored in the skies. She went and complained to Poseidon. Therefore, the constellations Ursa Major and Ursa Minor cross heaven, throughout the night without hiding into the seawater for a single moment. Even then Callisto's bliss would not diminish. Thus, she would spend eternity with her son.

It belongs to the Ursa Major family of the constellation. Its bordering constellations are Draco, Hercules, Ursa Major, Corona Borealis, Canes Venatici, Serpens, Coma Berenices, and Virgo. The major stars are- Arcturus (Alpha Bootis) is the third brightest star in the sky with an apparent magnitude of -0.04; Nekkar (Beta Bootis); Izar

(Epsilon Bootis); Muphrid (Eta Bootis); Alkaluros (Mu Bootis). The Notable Deep sky objects are Boötes void; Boötes I (Boötes Dwarf Galaxy); NGC 5466. It is associated with three meteor showers- January Bootids, June Bootids (June and July), Quadrantids (January). The best time to view the constellation is in June between +90 degree and -50 degree latitudes.

Caelum

Caelum, meaning the chisel, is the 81st largest constellation and the 7th smallest constellation in the sky. It only covers an area of 125 sq. Degree in the first quadrant of the southern hemisphere (SQ1).

It was introduced by the French astronomer Nicolas-Louis de Lacaille in the 18th century. It was first published in The southern star map of Lacaille in 1756. The original name of the constellation is named Caela Sculptoris. It was entitled after an instrument and an honor to the Age of Enlightenment. Lacaille depicted the constellation as a pair of crossed bruins connected by a ribbon.

It belongs to the Lacaille family of the constellation. The major stars of the constellations are Alpha Caeli, Gamma Caeli, X Caeli, and Beta Caeli. The neighboring constellations are Columba, Dorado, Eridanus, Horologium, Lepus, and Pictor. The notable deep-sky object in the Caelum is Seyfert Galaxy (HE0450-2958) and PN (G243-37). It is associated with one unnamed meteor shower. It is best visible in January between +40 degree and -90 degree latitudes in the sky.

Camelopardalis

Camelopardalis, the giraffe, is the 18th largest constellation in the sky. It covers an area of 757sq. degrees in the sky. It is located in the second quadrant of the northern hemisphere (NQ2). It was created by the Dutch astronomer Petrus Plancius and was documented by the German astronomer Jakob Bartsch in 1624. It was named after an animal giraffe, not a camel. Camelopardalis represents a giraffe with a long neck like a camel and has spots on the body like the leopard, and was called the "camel leopard".

Camelopardalis belongs to the Ursa Major family of the constellation. It is associated with the Camelopardalids meteor shower. The major stars in the constellation are Beta Camelopardalis, CS Camelopardalis, VZ Camelopardalis, Sigma 1694 Camelopardalis and Struve 1694 are binary stars. The neighboring constellations are Auriga, Cassiopeia, Cepheus, Draco, Lynx, Perseus, Ursa Major, and Ursa Minor. The best time to view the constellation is in February between +90 degrees and -10 degrees latitudes in the sky.

Cancer

Cancer is the 31st largest constellation located in the second quadrant of the northern hemisphere. It covers an area of 506sq. degrees. It is located in the second quadrant of the Northern Hemisphere (NQ2). The 2nd-century Greek astronomer Claudius Ptolemy discovered this constellation.

The constellation Cancer is also known as crab.

During the twelve tasks or labour of Hercules, one of them was to kill Hydra, a monster with many heads. During the aggressive combat, Hera sent a crab to increase an intrusion. The crab was ordered to attack Hercules. The crab squeezed on one of Hercules' toes with its claw. Out of exasperation, Hercules kicked it into the sky causing its death. The crab was placed in the sky as the well-known constellation Cancer, by Hera. None of its stars is bright enough because it did not succeed in its historic mission to kill Hercules.

It belongs to the Zodiac family of the constellation. The neighbouring constellations are Lynx, Leo Minor, Gemini, Leo, Canis Major, and Hydra. The major stars are- Acubens (Alpha Cancri), AI Tarf (Beta Cancri), Asellus Australis (Delta Cancri), Asellus Borealis (Gamma Cancri),

Tegmine (Zeta Cancri). Some notable deep sky objects are Beehive Cluster (NGC 2632), NGC 2775 (Caldwell 48), Messier 67 (NGC 2682). It is associated with the Delta Cancrids meteor shower. It is best visible in March between +90 degrees and -60 degrees latitudes in the sky.

Canes Venatici

Canes Venatici, the hunting dog, is the 38th largest constellation in the sky. It covers an area of 465sq. Degree in the third quadrant of the northern hemisphere (NQ3). It was introduced by the Polish astronomer Johannes Hevelius in 1678. It represents the two hunting dogs, Asterion and Chara, of its neighboring constellation Boötes. The 2nd century Greek astronomer Ptolemy included this constellation in Ursa Major which was later divided by Hevelius.

Canes Venatici belongs to the Ursa Major family of the constellation. It is associated with the Canes Venaticids meteor shower. The major stars of the constellations are- Cor Caroli (Alpha Canum), Chara (Beta Canum Venaticorum), La Superba (Gamma Canum Venatici), AM Canum Venaticorum and RS Canum Venaticorum. The bordering constellations are Boötes, Coma Berenice, and Ursa Major. Some notable deep sky objects are- Whirlpool Galaxy (NGC 5194), Sunflower Galaxy (NGC 5055), Whale Galaxy (NGC 4631), Cocoon Galaxy (NGC 4409), Hockey Stick Galaxies (NGC 4656 & NGC 4657), Messier 3 (NGC 5272). It is also associated with the Canes Venaticids Meteor shower.

Canis Major & Canis Minor

Canis Major, the greater dog, is the 43rd largest constellation in the sky. It is located in the first quadrant of the southern hemisphere (SQ1) and covers an area of 380sq. Degree in the sky.

It contains the brightest star in the sky- Sirius (Alpha Canis Majoris) with an apparent magnitude of -1.42; the other major stars are- Adhara (Epsilon Canis Majoris), Wezen (Delta Canis Majoris), Mirzam (Beta Canis Majoris), Aludra (Eta Canis Majoris). Some famous deep-sky objects are - Canis Major Dwarf Galaxy, Thor's Helmet (NGC 2359). The bordering constellations are Columba, Lepus, Monoceros, and Puppis.

Canis Minor, the lesser dog, is the 71st largest constellation in the sky as it covers an area of 183sq. degree in the sky. It is located in the second quadrant of the northern hemisphere (NQ2).

The major stars are- Procyon (Alpha Canis Minoris) and Gomeisa (Beta Canis Minoris). One of the famous Deep-sky objects in Canis Minor is NGC 2485. The neighbouring constellation is Cancer, Hydra, Monoceros, and Gemini. It is associated with the Canis-Minorids meteor shower.

Both the constellations were among the 48 constellations listed by the Greek astronomer Claudius Ptolemy in the 2nd century BC. They both belong to the Orion Family of the constellation.

Have you ever wondered why Canis Minor rises an hour before Canis Major? If yes, then you might also know the story behind that.

Canis Major, also termed as the big dog, was named Laelaps. He always caught whatever he hunted. He never missed his target. Once he was sent to hunt for the Teumessian fox, a fox that never could be caught. Right away, it was a matter of time. Everyone was eagerly waiting to know whether Laelaps was able to catch the Teumessian fox or the fox would be able to escape. Subsequently, Zeus turned the race into eternity. They both were turned into stone and placed into the sky by Zeus. The irony, of course, is they continue to chase in an eternal.

Capricornus

Capricornus, the sea-goat, is the 40th largest constellation located in the fourth quadrant of the southern hemisphere (SQ4). It covers an area of 414sq. Degree in the sky. It was listed by the 2nd-century Greek astronomer Claudius Ptolemy.

According to Greek mythology, Capricornus is depicted as the Greek God Pan, son of Hermes and Dryope, who was a one-half man and the other half goat. He is the god of wild, shepherds and flocks, nature of mountain wilds, rustic music and impromptus, and companion of the nymphs. He is regarded as the oldest Greek god associated with nature, wooden areas, and pasturelands. He always used to chase the nymphs and enhanced the part of the character of people born under this sign; this sign is responsible for a patient, ambitious and loyal character.

It belongs to the Zodiac Family of the constellation. The neighboring constellations are Aquarius, Aquila, Sagittarius, Microscopium, and Piscis Austrinus. The major stars are- Deneb Algedi (Delta Capricorni), Dabih (Beta Capricorni), Alshat (Nu Capricorni), Nashira (Gamma Capricorni), Dorsum (Theta Capricorni), Baten Algiedi (Omega Capricorni). It is associated with five meteor showers- Alpha Capricornids (July), Chi Capricornids,

Sigma Capricornids, Tau Capricornids, Capricornids Sagittarius. It is best visible in September between +60 and -90 degree latitudes in the sky.

Carina

Carina, the kneel, 34th largest constellation located in the second quadrant of the southern hemisphere (SQ2). In the 2nd century BC, the Greek astronomer Ptolemy cataloged a constellation called Argo Navis.

This constellation traditionally represented the ship Argo that Jason and the Argonaut used to sail to get the Golden Fleece. In the 18th century, the French astronomer Nicolas-Louis de Lacaille divided the constellation Argo Navis, the ship, because of its enormous size into three smaller parts Carina, the keel; Puppis, the stern, and Vela, the sails. The constellation Carina represents the body of the ship.

It belongs to the Heavenly Waters family of the constellation. The major stars of the constellations are Canopus- the second brightest star in the sky with an apparent magnitude of -0.74; Miaplacidus (Beta Carinae), Eta Carinae (Epsilon Carinae), Aspidiske (lota Carinae), and Omega Carinae. The neighboring constellations are Puppis, Vela, Centaurus, Chamaeleon, Musca, Pictor, and Volans. Some prominent deep-sky objects are Carina Nebula (NGC 3372); Southern Pleiades (Caldwell 102); Wishing Well Cluster (NGC 3532); The Diamond Cluster (Caldwell 96). It is associated with two meteor showers,

Alpha Carinids and Eta Carinids (January). The best time to view the constellation is in March between +20 degrees and -90 degrees latitudes in the sky.

Cepheus & Cassiopeia

Cepheus, the king, is the 27th largest constellation. It covers an area of 588 sq. Degree in the fourth quadrant of the Northern hemisphere (NQ4). It was listed by the 2nd-century Greek astronomer Claudius Ptolemy.

It belongs to the Perseus Family of the constellation. The major stars are- Alderamin (Alpha Cephei), Alfrik (Beta Cephei), Alrai (Gamma Cephei), Garnet Star (Mu Cephei); some famous deep-sky objects are the Fireworks Galaxy (NGC 6946), the Wizard Nebula (NGC 7380), the Iris Nebula (NGC 7023). The neighboring constellations are Cassiopeia, Camelopardalis, Cygnus, Draco, Lacerta, and Ursa Minor. It is best visible in March between +90 degrees and -10 degree latitudes.

Cassiopeia, the queen, is the 25th largest constellation as it covers an area of 598sq. Degree in the sky. It is located in the first quadrant of the Northern hemisphere (NQ1). It was cataloged by the 2nd-century Greek astronomer Claudius Ptolemy.

It belongs to the Perseus family of the constellation. Its neighboring constellations are Cepheus, Camelopardalis, Andromeda, Perseus, and Lacerta. The major stars are- Schedar (Alpha Cassiopeiae), Caph (Beta Cassiopeiae),

Ruchbah (Delta Cassiopeiae), Segin (Epsilon Cassiopeiae), Achird (Eta Cassiopeiae); some notable deep sky objects are the Pacman Nebula (NGC 281), the White Rose Cluster (NGC 7789), Messier 103 (NGC 581). It is also associated with the Perseids meteor shower in August. The best time to view the constellation in November between +90 degrees and -20 degrees latitude in the sky.

Cepheus was the king of Ethiopia and the father of the charming princess, Andromeda. Cassiopeia was the mother of Andromeda and the wife of King Cepheus. After saving Andromeda from the sea monster, Cetus; prince Perseus married her.

While celebrating the joy of a wedding, Phineas, the brother of Cepheus, turned against him and claimed that Andromeda had been promised to him first. Cepheus denied them and then they engaged in a furious fight. Perseus fought bravely against his opponents but he was outnumbered. He decided to use Medusa's head and turned his opponents into stone. Unfortunately, the king and the queen didn't look away from the head of Medusa and were turned into stone.

Centaurus

Centaurus, the centaur, is the 9th largest constellation as it occupies an area of 1060sq. degree in the sky. It is located in the third quadrant of the southern hemisphere (SQ3). It was among the 48 constellations listed by the 2nd century BC Greek astronomer Claudius Ptolemy.

Centaurus is associated with the Greek God Chiron, son of Titan King Cronus and Philyra. He was half man and half horse, a hybrid god. He was immortal and was famous for his wisdom and knowledge of medicine. He lived on Mount Pelion with his family consisting of his wife nymph Chariclo, son Carystus, and three daughters- Hippe, Endeis, and Ocyrhoe.

As was common, Chiron was working and from the sky, mid-way through the air an arrow gave and slew him. The arrow was released by Hercules to kill the fearful Hydra; it was poisoned with dreadful Hydra's blood, which has no cure. Chiron, being immortal, was suffering from intense pain. The only way to stop his unbearable pain was to place him among the stars as the constellation of Centaurus, as Zeus did.

It belongs to the Hercules Family of the constellation. The bordering constellations are Antlia, Vela, Carina, Musca,

Circinus, Crux, Hydra, Libra, Lupus. The major stars- Rigil Kentaurus (Alpha Centauri) is the fourth brightest star with an apparent magnitude of -0.27, Proxima Centauri (Alpha Centauri C) is the nearest star from our sun, Hadar (Beta Centauri), Muhlifain (Gamma Centauri), Ke Kwan (Kappa Centauri). The notable deep sky object is- Centaurus A (NGC 5128), Omega Centauri (NGC 5139), the Blue Planetary (NGC 3918).

It is also associated with three meteor showers- Alpha Centaurids (February), Omicron Centaurids (mid-February), and Theta Centaurids (February). It is best visible in the month of May between +25 degrees and -90 degrees latitudes in the sky.

Cetus

Cetus, the sea monster, is the 4th largest constellation as it occupies a large area of 1231sq. degree in the first quadrant of the Southern hemisphere (SQ1). It was listed by the 2nd-century Greek astronomer Claudius Ptolemy.

According to Greek Mythology, Cetus was a minion of the sea god, Poseidon. He was a ferocious, cruel sea monster or a whale. Poseidon ordered Cetus to devastate the city of Ethiopia when the sea nymphs complained to him that Cassiopeia, wife of Cepheus and mother of Andromeda, had boosted them.

Cetus was going soberly to destroy the city. Andromeda has chained to the rock as a heroic sacrifice to him. Perseus was going that way and got his eye struck at the lovely maiden chained to the rock. Perseus took out medusa's head and turned the enormous monster into a grey statue of rock.

Its bordering constellations are Aries, Pisces, Aquarius, Sculptor, Fornax, Eridanus, and Taurus. The major stars- Diphda (Beta Ceti), Menkar (Alpha Ceti), Mira (Omicron). Some notable deep sky objects are Messier 77 (NGC 1068), NGC 1055, NGC 1087, NGC 1042, NGC 247, NGC 47. It is also associated with three meteor showers- October Cetids, Eta Cetids, and Omicron Cetids. It is best visible in the month of November between +70 degrees to -90 degrees latitude in the sky.

Chamaeleon

Constellation Chamaeleon, the chameleon, is the 79th largest constellation located in the second quadrant of the southern hemisphere (SQ2). It occupies an area of 132sq. Degrees in the sky.

It was created by the Dutch astronomer Petrus Plancius in the 16th century. It was published in the Uranometria, the star atlas of Johann Bayer. It is named after a lizard that can it's colour. It depicts a Chameleon sticking its tongue to eat its neighbouring constellation Musca, the fly.

It belongs to the Johann Bayer family of the constellation. The major stars in the constellation are Alpha Chamaeleontis, Beta Chamaeleontis, Epsilon Chamaeleontis, R Chamaeleontis and CT Chamaeleontis. The neighbouring constellations are Apus, Carina, Musca, Mensa, Volans, and Octans. One of the famous deep-sky objects is Chamaeleon Cloud Complex. It is best visible in the month of April between +0 degree and -90 degree latitude in the sky.

Circinus

Circinus, the compass, is the 85th largest and 4th largest constellation located in the third quadrant of the Southern hemisphere (SQ3). It covers a small area of 93sq. Degree in the sky. It was introduced by the French astronomer Abbe Nicolas-Louis de Lacaille in 1756 and he depicted it as a compass. It represents the pair of dividing compasses used by the draughtsmen to measure the distance; it was titled after its shape for drawing circles. Its original name was Le Compas.

It belongs to the Lacaille family of the constellation. The bordering constellations are Centaurus, Musca, Apus, Norma, Lupus, and Triangulum Australe. The major stars of the constellations are Alpha Circini, Gamma Circini and Eta Circini. Some of the famous deep-sky objects are- Circinus Galaxy, Pismis 20, and Circinus X-1. It is associated with the Alpha Circinids meteor shower. It is best visible in the month of July between +30 degree to - 90 degree latitude in the sky.

Columba

Columba, the dove, is the 54th largest constellation in the sky. It is located in the first quadrant of the Southern hemisphere (SQ1). It covers an area of 270 sq. Degree in the sky. It was created by the Dutch astronomer Petrus Plancius in the 16th century and later it was introduced in Uranometria, Johann Bayer's star atlas. The original name of the constellation was Columba Noachi which means "Noah's dove." It was named after the biblical Dove that informed Noah about the Great receding flood. It represents the dove sent by Argonauts between the clashing rocks for its safe passage.

It belongs to the Heavenly Waters family of the constellation. The major stars of the constellation are Phact (Alpha Columbae), Wazn (Beta Columbae), Delta Columbae, Gamma Columbae, and Mu Columbae. Some of the famous deep-sky objects are- NGC 1792 and ESO 306-17. The neighboring constellations are Canis Major, Caelum, Lepus, Pictor, and Puppis. It is unassociated with any meteor shower. It is best visible in the month of February between +45 degrees to -90 degrees latitudes in the sky.

Coma Berenice

Coma Berenice, Berenice's hair, is the 42nd largest constellation in the sky. It is located in the third quadrant of the Northern Hemisphere (NQ3). It covers an area of 386sq. Degree in the sky. The Danish astronomer promoted this constellation in his star cataloged in 1602. It is named after queen Berenice II of Egypt.

Queen Berenice II of Egypt was the wife of Ptolemy III Evergetes. In 243 BC, his husband went on a risky mission against the Seleucids to avenge her sister's murder. During the third Syrian War, The queen swore to goddess Aphrodite she would cut her beautiful, long blonde hair if the goddess protects her husband in the war and he returns safely. Later, when her husband returned safely from the war, she went to the temple of Aphrodite and cut off her hair. But her hair disappeared the next which became a reason for the king's anger. To please the king, Berenice offended the court astronomer to extricate her for the angered king. When the court astronomer Conon was called, he pointed towards a group of the star and said the goddess had placed her hairs in the sky.

This constellation belongs to the Ursa Major family of the constellation. The bordering constellations are Boötes, Canes Venatici, Leo, Ursa Major, and Virgo. The 4 major stars of the constellation are Diadem (Alpha Comae Berenices), Beta Comae Berenices, Gamma Comae Berenices and FK Comae Berenices. The most notable deep-sky object in the constellation is the Black Eye Galaxy (M64, NGC 4826) is an unusual spiral galaxy also called the Sleeping Beauty Galaxy or Evil Eye Galaxy. Other than that, Coma Star Cluster (Melotte 111), Coma Cluster of galaxies, and Virgo Cluster of galaxies.

It is associated with the Coma Berenicids meteor shower in the month of December. It is best visible in the month of May between +90 degree and -70 degree latitudes in the sky.

Corona Australis

Corona Australis, the southern crown, is the 80th largest constellation located in the third quadrant of the Southern Hemisphere (SQ3). It covers an area of 128sq. Degree. It was cataloged by the Greek astronomer Claudius Ptolemy in the 2nd century BC. It belongs to the Hercules family of the constellation.

Semele was the charming daughter of the king of the thieves and the lover of Zeus. Zeus made an appearance of the immortal and never refused that he was an Olympian god. The demure mortal, Semele was pregnant, unaware of the truth that his lover is the supreme god of sky and thunder.

One gratifying morning, Semele saw that his lover vacated the bed a few minutes ago. She rises from her bed and asks Maybury to fit her out. The Maybury no other than Hera, wife of Zeus. Hera placed the seed of Skepticism in the soul of Semele. That sulky evening when Zeus returned, Semele questioned Zeus to examine his real identity. She requested to see the all mighty supreme god. Zeus was unable to deny her lover the supreme god revealed his marvelous form to her. By the time Semele's skin started to burn and slowly quash her body. From her ashes came out a little divine boy named Dionysus. Dionysus, the god of wine, grew up and fought with Hades to recover his mother's soul. Her mother's soul was finally recovered and came back to Mount Olympus. Dionysus placed her mother's eternal life in the sky. Corona Australis represents her crown.

It belongs to the Hercules family of the constellation. The bordering constellations are Sagittarius, Scorpius, Ara, and Telescopium. The major stars are- Meridiana (Alpha Coronae Australis), Beta Coronae Australids, and RX J1856.5-3754. Famous deep-sky objects are Corona Australis Nebula and Coronet Cluster. It is associated with the Corona Australids meteor shower. The best time to view the constellation is in the month of August between +40 degrees and -90 degrees in the sky.

Corona Borealis

Corona Borealis, the northern crown, is the 73rd largest constellation in the sky. It covers an area of 179sq. Degree in the third quadrant of the Northern Hemisphere (NQ3). It was described by the Greek astronomer Claudius Ptolemy in the 2nd century BC.

Ariadne was the sympathetic daughter of the King Minos of Crete, son of Zeus and Europa. King Minos kept a strong, ferocious flesh-eating monster with a bull's head and a human body named Minotaur in a Babyrinth. The king ordered seven boys and seven girls to be brought every year for the Minotaur to eat on the island of Crete.

Theseus, the Athenian king and the son of Poseidon, agreed to be one of the seven sacrificed men and set sail for the island of Crete. Ariadne mucked in with Theseus to defeat the monster. She presented him with a sword to cleave the monster and a thread to find the way to get out of the Babyrinth. In the barbarous battlefield, Theseus put Minotaur to the sword and found his way out of the maze. In reward, Theseus married Ariadne and took her with him to Athens. However, he didn't love her, and he abandoned her on the island of Xaros. Spending her lonesome and regretful days there on the island, she met Dionysus, son of Zeus and Semele. She found her true love and got married. Then, she died leaving Dionysus in an agonizing state. He placed her crown Ariadne in the sky where the stars became the jewels.

It belongs to the Ursa Major family of the constellation. The bordering constellations are Boötes, Hercules, and Serpens Caput. The major stars of the constellations are- Alphecca (Alpha Coronae Borealis), Nuskan (Beta Coronae Borealis), Zeta Coronae Borealis, and Struve 1967 (Gamma Coronae Borealis). One of the famous deep-sky objects is- Corona Borealis Galaxy Cluster (Abell 2065). It is best visible in the month of July between +90 degree to -50 degree latitude in the sky.

Corvus and Crater

Corvus, the crow, is the 70th largest constellation located in the third quadrant of the Southern Hemisphere sky, covering an area of 184sq. Degree. It was cataloged by the Greek astronomer Claudius Ptolemy in the 2nd century BC. It belongs to the Hercules Family of the constellation.

Crater, the cup, is the 53rd largest constellation in the sky. It covers an area of 282sq. Degree in the second quadrant of the Southern hemisphere (SQ3). It was cataloged by the Greek astronomer Claudius Ptolemy in the 2nd century BC. It belongs to the Hercules family of the constellation.

Corvus was the sacred crow of Apollo, son of Zeus and Leto, and the twin brother of Artemis. Apollo was the god of music, poetry, art, prophecy, truth, archery, and everything.

One sacred day, Apollo was about to perform a sacrifice to Zeus for which he required water. Consequently, he ordered Corvus to fetch the water from a close-by spring and gave him a cup-crater. Corvus starts his way to fetch water but soon stumbles across a huge mouth-watering fig tree. He gazed up at the figs and decided to wait until the fig ripens. This takes a few days, and the poor crow realizes the importance of time. Then he tried finding an excuse to explain his absence. Anxious Corvus finds a way to vindicate himself. He planned that water snake Hydra kept him from completing his task promptly. So, with the Hydra in one claw and the crater in the other, he flies back to Apollo. Apollo being the god of truth and prophecy identifies the real reasons for foolish birds. He grabs the crow and throws it in the sky out of frustration and ostracizes the three to the sky for eternity.

Crux

Crux, the southern cross, is the smallest constellation in the sky. It covers a small area of 68sq. Degree. It is located in the third quadrant of the southern hemisphere. It represents a cross-shaped asterism that is known as the Southern Cross. It was created by the Dutch astronomer Petrus Plancius in 1613. Earlier it was part of the constellation Centaurus. In 1679 a French astronomer Augustin Royer separated Crux from Centaurus. In the 2nd century Ptolemy, a Greek astronomer, listed the constellation as a part of Centaurus. By 400AD, Crux was not visible from the Northern Hemisphere. Currently, it is only visible from the circumpolar regions like Australia and Zealand.

It belongs to the Hercules Family of the constellation. The major stars of the constellation are Acrux (Alpha Crucis), Mimosa (Beta Crucis), Gracrux (Gamma Crucis), and Delta Crucis. It is associated with the Crucids meteor shower. The neighboring constellations are Centaurus and Musca. The famous deep-sky objects are- Coalsack Nebula (Caldwell 99) and the Jewels Box (Caldwell 4755). It is best visible in the month of May.

Cygnus

Cygnus, the swan, is the 16th largest constellation located in the fourth quadrant of the Northern Hemisphere (NQ4). It covers an area of 804sq. Degree in the sky. It belongs to the 48 constellations listed by the 2nd Century BC Greek astronomer Ptolemy.

Cygnus was an elegant swan.

Its vision was enough to capture the most desired heart. Cygnus was no other than Zeus in disguise. Zeus had transformed himself into a beautiful swan to seduce Leda.

Leda, the wife of the Spartan king, Tyndareus, and Helen of Troy possessed a supreme beauty. No one could escape from her enchantress. Zeus was no exception. He desired her love, but at the time, he knew that he wouldn't be able to obtain Leda's love. Therefore, he transformed himself into an elegant swan and won her love. Pollux was born from their union.

It belongs to the Hercules Family of the constellation. The bordering constellations are- Cepheus, Draco, Lacerta, Lyra, Pegasus, and Vulpecula. The major stars are- Deneb (Alpha Cygni), Sadr (Gamma Cygni), Aljanah (Epsilon Cygni), Fawari (Delta Cygni), and Albireo (Beta Cygni).

Some of the famous deep-sky objects are- Fireworks Galaxy (NGC 6946), North America Nebula (NGC 7000), Pelican Nebula (IC 5070), Crescent Nebula (NGC 6888), and Cygnus Loop. It is associated with two meteor showers- October Cygnids and Kappa Cygnids (August month). It is best visible in the month of September.

Delphinus

Delphinus, the dolphin, is the 69th largest constellation located in the fourth quadrant of the Northern hemisphere (NQ4) and occupies an area of 189sq. Degree in the sky. It was described by Ptolemy in the 2nd century BC. It belongs to the Heavenly Waters Family of the constellation.

Arion was a well-known skilled singer and poet. He used to play the harp and sing the song in a melodious soothing voice. He was an extremely famous singer in Greece during the 7th century BC. He used to give his performance in different countries throughout the Mediterranean.

Once he was enjoying the pleasant journey from Southern Italy to Greece, the sailor on the ships colluded themselves against him. They planned to kill and grab all his earnings; and started to surround him. Arion finding himself surrounded begged the sailor to allow him to sing one last song. As he began to sing with the allowance, The dolphins were attracted towards the ship. He played his harps and sang wonderfully mellifluously. Several dolphins swam alongside the ship, Arion somehow managed to jump out of the ship and the dolphins saved Arion and carried him back to Greece. Apollo, the god of music, replaced the dolphins near the constellation of Lyra.

It belongs to the Heavenly Waters Family of the constellation. The bordering constellations are Pegasus, Aquarius, Aquila, Sagitta, Equuleus, and Vulpecula. The major stars are- Rotanev (Beta Delphini), Sualocin (Alpha Delphini), Adulfin (Epsilon Delphini), and Musica (18 Delphini). One of the famous deep-sky objects is the Blue Flash Nebula (NGC 6905) a small planetary nebula discovered by William Herschel. It is not associated with any meteor shower. The best time to view the constellation is in the month of September.

Dorado

Dorado, representing the swordfish in Spanish, is the 72nd largest constellation as it covers an area of 179sq. Degree in the sky. It is located in the first quadrant of the Southern hemisphere (SQ1). It is a small and faint constellation. It was created by the Dutch astronomer Petrus Plancius. It was first published in the Uranometria, star atlas of Johann Bayer in 1603.

It belongs to the Johann Bayer family of the constellation. The neighboring constellations are Caleum, Horologium, Reticulum, Hydrus, Mensa, Volans, and Pictor. The major stars of the constellations are Alpha Doradus, Beta Doradus, Gamma Doradus, Delta Doradus, HD 29712, and S Doradus. The famous deep-sky object Large Magellanic Clouds (LMC) is an irregular galaxy located in the constellation. Some other deep-sky objects are- Tarantula Nebula (NGC 2070), and the Ghost Head Nebula (NGC 2080). It is not associated with any meteor shower and is best visible in the month of January.

Draco

Draco, the dragon, is the 17th largest constellation located in the circumpolar region. It covers an area of 1083sq. Degree and is located in the third quadrant of the Northern hemisphere (NQ3). It is one of the 48 constellations discovered by Ptolemy.

It represents the mythical dragon Ladon.

Ladon was a great dragon and the guard of the golden Apple tree in the garden of Hesperides.

Gaia gave Hera a golden Appletree as a wedding present when she married Zeus. Hera placed the sacred Apple tree in the garden of Mount Atlas and tasked Atlas's daughter, the Hesperit, with guarding it. She also placed the dragon Ladon as the guard to prevent Hesperides or anyone else from picking up the apples.
It belongs to the Ursa Major Family of the constellation. The bordering constellations are- Boötes, Camelopardalis, Cepheus, Cygnus, Hercules, Lyra, Ursa Major, and Ursa Minor. The major stars of the constellations are Eltanin (Gamma Draconis), Aldibain (Eta Draconis), Aldibain (Beta Draconis), Altais (Delta Draconis), Aldhibah (Zeta Draconis), Edasich (Iota Draconis), Thuban (Alpha Draconis). The famous deep-sky objects are- Cat's Eye

Nebula (NGC 6543), Spindle Galaxy (NGC 5866), Draco Dwarf Galaxy, and Tadpole Galaxy. It is associated with the Draconids meteor shower which occurs in October. It is best visible in July.

Equuleus

Equuleus, the little horse, is the 2nd smallest constellation in the sky listed by the Greek astronomer Claudius Ptolemy in the 2nd century BCE. It covers an area of 72sq. Degree in the fourth quadrant of the Northern hemisphere (NQ4).

Hippe was the dutiful daughter of the Centaur Chiron. She was seduced by Aeolus and became pregnant with a child. But she was terribly ashamed to inform her father that she was going to give birth to Aeolus' child. Hence, she decided to hide herself and escaped to the mighty mountains, and gave birth to the child, Melanippe. When his father was searching for her daughter, he came to the mountains; Hippe prayed to the gods to hide from her father. Goddess Artemis changed Hippe into a mare and placed it in the boundless sky behind the Pegasus.

It belongs to the Heavenly Waters family of the constellation. The bordering constellations are- Aquarius, Delphinus, and Pegasus. The major stars are Kitalpha (Alpha Equulei), Delta Equulei, and Gamma Equulei. This constellation does not have any famous deep-sky object. It is best visible in the month of September.

Eridanus

Eridanus, the river, is the 6th largest constellation located in the first quadrant of the Southern hemisphere (SQ1). It covers an area of 1138sq. Degrees in the sky. It was listed by the 2nd Century Greek Astronomer Ptolemy and was among the list of 48 constellations.

The story of Eridanus is associated with Phaethon, son of Helios.

One day Phaethon sought his father to allow him to drive the maleficent chariot of the sun. His father hesitantly allowed him and provided him the hold sway. Due to the lack of experience, Phaethon lost control of the chariot and the way the sun-chariot swerved in the wrong direction as a result it was on fire. The plain fertile land of Africa turned into the barren desert and had become the reason for the black color of men. Zeus, exasperated by the boy's mistake, punishes him with a thunderbolt. He hurls his burning body in river Eridanus. After his death, he was placed among the stars as the constellation of Eridanus.

It belongs to the Heavenly Waters family of the constellations. Its neighboring constellations are Caelum, Cetus, Fornax, Horologium, Hydrus, Lepus, Orion,

Phoenix, Taurus, and Tucana. The major stars are-Achernar (Alpha Eridani), Cursa (Beta Eridani),Acamar (Theta Eridani), Ran (Epsilon Eridani), and Zaurak (Gamma Eridani). Some famous deep-sky objects are-Witch Head Nebula (IC 2118), Eridanus Supervoid, and Eridanus Group. It is best visible in the month of December.

Fornax

Fornax, the furnace, is the 41st largest constellation located in the first quadrant of the Southern Hemisphere (SQ1). It covers an area of 398sq. Degrees in the sky. It was introduced by the French astronomer Nicolas-Louis de Lacaille in 1756. It was described as the chemical Furnace with an alembic and receiver. The original name of the constellation is Fornax Chemica, which was shortened to Fornax by the English astronomer Francis Bailey in 1845.

It belongs to the Lacaille family of the constellation. The major stars of the constellation are Dalim, Beta Fornacis, HD 16417, HIP 13044, HD 20781, and HD 20782. It is bordered by Cetus, Sculptor, Phoenix, and Eridanus. The famous deep-sky objects are- Great Barred Spiral Galaxy and Fornax Dwarf. It is best visible in the month of December.

Gemini

Gemini, the twins, Castor and Pollux, is the 30th largest constellation in the sky. It is located in the second quadrant of the Northern hemisphere (NQ2) and covers an area of 514Sq. Degree. It is described by the Greek astronomer Claudius Ptolemy in the 2nd Century AD and the list of Ptolemy's 48 constellations. It depicted the twins, Castor and Pollux.

Gemini is the story of identical twins.

Leda was the gorgeous wife of the Spartan king Tyndareus. She was very beautiful and entered the eye of Zeus. The great god of Mount Olympus, Zeus transformed himself into Swan and seduced Leda. Leda gave birth to the two sons, the twin's Pollux and Castor; together they were called Dioscuri and she also gave birth to the twin sisters, The Helen of Troy and Clytemnestra. Pollux (Polydeuces) was the immortal son of Zeus and Leda; Castor was the son of Leda and Tyndareus. He was a mortal son of Leda, therefore he couldn't be alive forever.

They were great on the battlefield full of energy, cleverness, and imagination. Pollux was a great fighter whereas Castor was a great horseman. Together they went with Jason on the Argo Navis on the adventure to find the

Golden Fleece, and they also saved the ship from a terrible storm. They were called the Patron saints of the sailor. The twins were engaged with the sisters, Phoebe and Hilaira.

After coming back from the adventure, their sister Helen was in love with the prince of Troy and left the Spartan family. This outraged the Great Spartan king and began the war to bring back her daughter from the Troys. The mighty army was set, and Dioscuri along with Jason and his Argonauts joined the fierce battle to bring back his sister. This was termed the Trojan War. Unfortunately, in the war, Castor was killed. Pollux requested his father Zeus to bring him back. Zeus agreed to his dear son and gave him two options either to let his brother die or to give him half of his immortality. Pollux can not bear his life without his dear brother and choose to share half of his immortality. Zeus placed the twins in the sky as the constellation of Gemini. Zeus also rewarded them with the Golden Fleece and immortalized the Ram in the sky as the constellation Aries.

It belongs to the Zodiac Family of the constellation. It is associated with two meteor showers namely Geminids and Rho Geminids. The brightest star of the constellation is Pollux, next to it is Castor and then Alhena, Mebsuta, Tejat Posterior, Tejat Prior, and Alzirr. It is bordered by Auriga, Orion, Canis Major, Cancer, Monoceros, Lynx, and Taurus. The constellation contains the famous open star cluster called Messier 35 (NGC 2168) and the Eskimo Nebula (NGC 2392, Caldwell 39); the Jellyfish Nebula (IC 443), and the Medusa Nebula (Abell 21). It is best visible in the month of February.

Grus

Grus, representing the crane, is the 45th largest constellation and occupies an area of 366sq. Degree in the sky. It is located in the fourth quadrant of the Southern hemisphere (SQ4). It was created by the Dutch astronomer Petrus Plancius. It was first published in the Uranometria, a star atlas of Johann Bayer. Grus is the sacred bird to Hermes.

It belongs to the Johann Bayer family of the constellation. The major stars of the constellation are Alnair, Gruid, Al Dhanab, Delta Gruis, and Gliese 832. It is bordered by Indus, Microscopium, Pisces Austrinus, Sculptor, Phoenix, and Tucana. The major stars are- Alnair (Alpha Gruis), Gruid (Beta Gruis), Al Dhanab (Gamma Gruis) and Gliese 832. Some famous deep-sky objects are the Grus Quartet, a group of four interacting galaxies; Spare Tyre Nebula (IC 5148). It is best visible in the month of October.

Hercules

Hercules, named after Heracles, is the 5th largest constellation in the third quadrant of the Northern hemisphere (NQ3). It covers an area of 1225sq. Degree in the sky. It was catalogued by the Greek astronomer Ptolemy in the 2nd century BC. It represents the penultimate labour of Heracles.

Hercules was born to Zeus, the king of the gods, and Alcmene, a mortal woman. Zeus had a lot of offspring in external affairs, but his wife Hera hated Heracles (Hercules) the most. Zeus had disguised as the husband of Alcmene and had Hercules. The hate of Hera on Hercules made him mad, and he killed all his children born through his first wife Megara. He did a dreadful crime and wanted to atone. He was first cured of his madness by Antikyreus which made him realize what he had done. Then he fled to the Oracle of Delphi. The Oracle, then guided by Hera, directed Hercules to the king Eurystheus who, also guided by Hera, gave him twelve tasks.

These twelve tasks tested his strength and made him famous as a great hero and known as the Twelve Labours of Heracles (Hercules).

- In the first task, he killed the ferocious Nemean Lion

with direct hands.

• In the second task, he killed Hydra that was a nine-headed dragon sacred to Hera.

• In the third labour, he captured the Cerynerian Hind. It was said that it could run faster than arrows and thus hard to capture!

• In the fourth labour, Heracles (Hercules) captured the Erymanthian Boar. He captured it with a plan in mid-winter in snow.

• In the fifth labour, he had to clean Augean stables. As the livestock was divinely healthy, they constantly filled with dung. It was hard to finish and very humiliating. But Hercules did it.

• In the sixth labour, Hercules killed the Stymphalian birds, which were man-eating birds and had wings of brass.

• In the seventh labour, Hercules captured the Cretan Bull which was responsible for the birth of the Minotaur.

• In the eighth labour, Hercules efficiently captured the man-eating horses called the Mares of Thrace.

• In the ninth task, Hercules promptly stole the magical girdle of the queen of Amazons of female warriors. It was an easy task as Queen Hippolyte was intrigued by Hercules's masculinity and gave him the girdle without a fight.

• In the tenth labour, Heracles captured the cattle of Geryon.

• In the eleventh task, he had to bring the apples of the Hesperides.

• In the last and twelfth task, Hercules had captured the multi-headed hound Cerberus which was guarding the underworld.

All these labours were amazing tales of Hercules' strength and heroism.

It belongs to the Hercules Family of the constellation. The bordering constellations are Aquila, Boötes, Corona Borealis, Draco, Lyra, Ophiuchus, Sagitta, Serpens Caput, and Vulpecula. The major stars are- Kornephoros (Beta Herculis), Sarin (Delta Herculis), Rasalgethi (Alpha Herculis), Sophian (Eta Herculis), and Rukbalgethi Genubi (Theta Herculis). Some famous deep-sky objects are- The Great Globular Cluster (NGC 6205), Hercules Cluster (Abell 2151), and Hercules A. It is associated with the Tau Herculids meteor shower which occurs in the month of May. It is best visible in the month of July.

Hydra

Hydra, the sea serpent, is the largest constellation in the sky. It occupies an area of 1303 sq. degrees in the second quadrant of the Southern hemisphere (SQ2). It was listed in the list of 48 constellations cataloged by the 2nd-century Greek astronomer Claudius Ptolemy. It represents a water snake.
Hydra, one of the famous beast which was killed by Hercules in its 12 labour. Hydra was also known as Lernean Hydra.

Hydra was an offspring of Typhon and Echidna, who was half woman and half snake. Together, they provided Hydra their immortality. Hydra was immortal and had revival abilities. He had one immortal head, which was covered and protected by other lethal heads that grew around it. Hydra was a monstrous shape and had an iniquitous disposition. His blood was super toxic and was almost an undefeatable beast. Hera was the wife of Zeus. She adopted Hydra intending to destroy Hercules. She raised the fierce beast and gave him extraordinary powers.

Hercules went to the Lernaean Swamp with its mouth and nose covered with a thick fabric that would prevent him from breathing the poisonous scent of the fierce beast. When the fight began, Hercules started cutting Hydra's head as soon as possible. After a few minutes of bloody battle, Hercules realized it was almost impossible to defeat Hydra alone. Desperate, he called his nephew Lolaus, for help. This frustrated Hera and she sent a jumbo crab to distract Hercules. He somehow managed to reach the immortal head of Hydra and snapped it with a golden sword given by Athena. Hercules placed the dead beast under a giant rock. Although, Hydra failed in his mission to kill Hercules. Hera placed him in the sky among the stars for his bravery.

It belongs to the Hercules Family of the constellation. Its bordering constellations are- Antlia, Cancer, Canis Major, Centaurus, Corvus, Crater, Leo, Libra, Lupus, Monoceros, Puppis, Pyxis, Sextans, and Virgo. The major stars are- Alphard (Alpha Hydrae), Minchir (Sigma Hydrae), and Zeta Hydrae. Some famous deep-sky objects in the constellations are Hydra Cluster, Southern Pinwheel Galaxy, Tombaugh's Globular Galaxy, and Ghost of Jupiter (Caldwell 59) a planetary nebula. It is associated with two meteor showers- Alpha Hydrids and Sigma Hydrids. It is best visible in April.

Hydrus

Hydrus, the water snake, is the 61st largest constellation as it occupies 243sq. Degree in the sky. It is located in the first quadrant of the Southern Hemisphere (SQ1). It is called the male water snake and is also known as the lesser water snake. It was created by the Dutch astronomer Petrus Plancius in 1597. It represents the sea. Hydrus was the counterpart of the largest constellation Hydra. In 1756, the French astronomer Nicolas-Louis de Lacaille distinguished these two constellations.

It belongs to the Johann Bayer family of the constellation. The major stars of the constellation are Beta Hydri, Alpha Hydri, Gamma Hydri, Delt Hydri, and Epsilon Hydri. Some major deep sky objects are IC 1717, and NGC 1511. The neighboring constellations are Dorado, Reticulum, Horologium, Eridanus, Mensa, Octans, Phoenix, and Tucana. It is best visible in November.

Indus

Indus, India, is the 49th largest constellation located in the fourth quadrant of the northern hemisphere. It occupies an area of 294sq. Degrees in the sky. It was created by Dutch astronomer Petrus Plancius in 1603. He depicts the constellation as a naked man holding arrows in both his hands. It was first depicted in the Uranometria, a star atlas of Johann Bayer. In Uranometria, the constellation depicts a native of Madagascar.

It belongs to the John Bayer family of the constellation. The major stars of the constellation are the Persian (Alpha Indi), Beta Indi, Epsilon Indi, Theta Indi, and Rho Indi. Some famous deep-sky objects are NGC 7064, NGC 7140, and IC 5152. It is bordered by Sagittarius, Grus, Octans, Pavo, Telescopium, and Tucana. It is best visible in September.

Lacerta

Lacerta, the lizard, is the 68th largest constellation located in the fourth quadrant of the Northern hemisphere. It was created by Polish astronomer Johannes Hevelius in 1687. It is also called Little Cassiopeia because of its W shape. Hevelius introduced the constellation in his atlas named Firmamentum Sobiescianum. It represents a lizard.

It belongs to the Perseus family of the constellation. The major stars of the constellation are Alpha Lacertae, Beta Lacertae, EV Lacertae, Roe 47, and ADS 16402. Some famous deep-sky are NGC 7243 (Caldwell 16) and BL Lacertae. The neighboring constellation is Andromeda, Cassiopeia, Cepheus, Cygnus, and Pegasus. It is best visible in October.

Leo

Leo, the lion is the 12th largest constellation as it covers an area of 947sq. Degree in the second quadrant of the Northern hemisphere (NQ2). It was among the 48 constellations listed by the Greek astronomer Claudius Ptolemy in the 2nd century.

Leo was better known as the Nemean lion. The Nemean lion was a fearsome beast who had terrorized the land of Nemea and killed all the people who ventured near it. In the twelve labor of Hercules, the killing of the Nemean lion was the first task. His skin was so insensitive and stocky that it was impervious to metals. His claws and fangs were like iron. Because of his impenetrable skin, Hercules wrestled it with bare hands and finally managed to strangle the beast. Hercules removed his skin and wore it and also used its head as a protective helmet. The Nemean lion was placed in the sky as the constellation Leo.

It belongs to the Zodiac Family of the constellation. The neighboring constellations are- Cancer, Coma Berenices, Crater, Hydra, Leo Minor, Lynx, Sextans, Ursa Major, and Virgo. The major stars are- Regulus (Alpha Leonis), Denebola (Beta Leonis), Algieba (Gamma Leonis), Zosma (Delta Leonis), Chort (Theta Leonis), Adhafera (Zeta

Leonis), and Wolf 359. The most notable deep-sky object is the Leo Ring. It is associated with the Leonids meteor shower in November. This constellation is best visible in April.

Leo Minor

Leo Minor, the smaller lion, is the 64th largest constellation. It is located in the second quadrant of the Northern hemisphere (NQ2). It was created by the Polish astronomer Johannes Hevelius in 1687. It was named by an English astronomer Richard A.Proctor. He changed the name of this constellation from Leaena to Leones.

It belongs to the Ursa Major family of the constellation. It is associated with Leonis Minorids. It is bordered by Cancer, Leo, Lynx, and Ursa Major. The major stars of the constellation are Praecipua- 46 Leonis Minoris, Beta Leonis Minoris, 21 Leonis Minoris, 10 Leonis Minoris, and 37 Leonis Minoris. The notable deep-sky object in this constellation is Hanny's Voorwerp, it is a quasar. It is also associated with the Leonis Minorids meteor shower that occurs during October month. It is best visible in April.

Lepus

Lepus, the hare, is the 51st largest constellation as it covers an area of 290sq. Degree in the second quadrant of the Northern Hemisphere (NQ2). It was cataloged by the 2nd century BC Greek astronomer Ptolemy. It represents the hare.

Lepus does not have any particular Greek mythology associated with it. It is represented as a hare being chased by the great hunter Orion or by his hunting dogs, represented by the constellation Canis Major and Canis Minor.

It belongs to the Orion Family of the constellation. The bordering constellation is Orion, Caelum, Canis Major, Columba, Eridanus, and Monoceros. The major stars are- Arneb (Alpha Leporis), Nihal (Beta Leporis), Hind's Crimson Star (R Leporis), and Gliese 229. One of the notable deep-sky objects is- Spirograph Nebula (IC 418). It is not associated with any meteor shower. This constellation is best visible in January.

Libra

Libra, the balance, is the 29th largest constellation located in the third quadrant of the Southern Hemisphere (SQ3). It covers an area of 538sq. Degrees in the sky. It is listed by the 2nd-century Greek astronomer Ptolemy. It represents a scale.

In Greek Mythology, Themis, the Titan goddess represents justice and balance, harmony and equilibrium, and is called the Goddess of Justice. She was the daughter of the Titan king Uranus and the earth god, Gaia, and the wife of Zeus. She was depicted holding a pair of scales in one hand and a sword in the other and showing that Justice is Blind; she is considered as the Lady Justice.

It belongs to the Zodiac Family of the constellation. The neighboring constellations are Virgo, Hydra, Scorpius, Centaurus, Lupus, Ophiuchus, and Serpens Caput. The major stars of the constellations are Zubeneschamali (Beta Librae), Zubenelgenubi (Alpha Librae), Brachium (Sigma Librae), Zubenelakrab (Gamma Librae), Zuben Elakribi (Delta Librae), and Gliese 581 (HO Librae). Libra contains two barred spiral galaxies- NGC 5885 and NGC 5792; NGC 5897 is a globular cluster and NGC 5890 is an unbarred lenticular galaxy. It is also associated with the

May Librids Meteor shower. The best time to view the constellation is the month of June.

Lupus

Lupus, the wolf, is the 46th largest constellation located in the third quadrant of the Southern Hemisphere (SQ3). It was among the list of 48 constellations listed by the Greek astronomer Ptolemy in the 2nd century BC.

Lupus used to be part of Centaurus. It represented a hybrid creature with a human head and torso and the legs and tail of the lion. It is called Therium and is considered a wild beast.

It belongs to the Hercules Family of the constellation. Its bordering constellations are Hydra, Scorpius, Libra, Centaurus, Circinus, and Norma. The major stars are- Alpha Lupi, Beta Lupi, Gamma Lupi, Delta Lupi, and Lupus-TR-3. One of the notable deep-sky objects located is the Retina Nebula (IC 4406). It is not associated with any meteor shower. The best time to view the constellation is in June.

Lynx

Lynx, the tiger, is the 28th largest constellation in the sky. It is located in the second quadrant of the Northern Hemisphere. It was introduced by the Polish astronomer Johannes Hevelius in the 17th century. It was named after an animal, a tiger. A sailor named Lynceus, who sailed with Jason and the Argonauts, was said to have extremely keen eyesight. It was even said he could see things underground. He was part of the expedition to recover the legendary Golden Fleece. It is a faint constellation in the northern hemisphere between Auriga and Ursa Major.

This constellation Belongs to the Ursa Major family of the constellation. It contained five named stars. The major stars of the constellation are Alpha Lyncis, Alsciaukat (31 Lyncis), 38 Lyncis, 12 Lyncis, 19 Lyncis, and 6 Lyncis. The notable deep-sky objects are- NGC 2419 (Caldwell 25) is a globular star cluster; UFO Galaxy (NGC 283), and Bear's Paw Galaxy (NGC 2537). It is bordered by Auriga, Camelopardalis, Cancer, Gemini, Ursa Major, Leo, and Leo Minor. It is associated with two meteor showers- Alpha Lyncids and September Lyncids. The best time to view the constellation is in March.

Lyra

Lyra, the lyre, is the 52nd largest constellation in the sky. It is located in the fourth quadrant of the Northern hemisphere (NQ4). It occupies an area of 286sq. Degrees in the sky. It was among the 48 constellations listed by Claudius Ptolemy in the 2nd century BC. It represents a musical instrument, the lyre.

Orpheus, the son of the Thracian King and Muse Calliope, was a talented musician. He was a brilliant musician; he used to play the lyre very charmingly. Everyone was attracted by his melodious music. Even the god of music Apollo was impressed by his skills and awarded him a harp. The divine beautiful harp was made by Artemis, Apollo's sister.

Soon he fell in love with an arresting maiden Eurydice and decided to marry. On the day of the marriage, she was trying to escape from satyr and fell in the nest of vipers; she died leaving a mournful bias in Orpheus' mind.

After the sadful moment, he got advice to bring Eurydice back to life from the underworld. On getting a single chance to bring his love back, he lined up the way to hell. He met Hades, the god of the underworld, requested him to free Eurydice, and started playing his music. The dulcet music changed the scenario. Hades freed Eurydice on the condition that while going back to the upper world Orpheus should be in the front and not look behind. They moved under the condition; Orpheus didn't look back and went the way back to land. When Orpheus reached the upper world, the voice of Eurydice's footsteps was missing, so he looked back and everything was safe, Eurydice was there for a while but soon she disappeared. Maenads killed Orpheus and his lyre was carried by the Zeus' eagle to Greece, where he placed it in the sky.

It belongs to the Hercules Family of the constellation. The bordering constellations are Cygnus, Draco, Hercules, and Vulpecula. The brightest star in the constellation is- Vega (Alpha Lyrae) with an apparent magnitude of 0.03 and is the 4th brightest star in the sky. The other stars are- Sulafat (Gamma Lyrae), Sheliak (Beta Lyrae), Alathfar (Mu Lyrae), and Gliese 758. The notable deep-sky objects are- the Ring Nebula (NGC 6720), NGC 6779, NGC 6791, NGC 6745, and IC 1296. It is associated with 3 meteor showers namely, Lyrids, June Lyrids, and Alpha Lyrids. The best time to view the constellation is in August.

Microscopium

Microscopium, the microscope, is the 66th largest constellation located in the fourth quadrant of the Southern hemisphere (SQ4). It was introduced by the French astronomer Nicolas-Louis de Lacaille. It occupies an area of 209.5 sq. degrees in the sky.

It represents the microscope, an instrument that enlarges the image of small objects. It is named after the compound microscope invented by Zacharias Janssen that uses a lens. The constellation depicts a tube above a square box.

It belongs to the Lacaille family of the constellation. The 5 major stars of the constellation are Gamma Microscopii, Epsilon Microscopii, Theta Microscopii, Alpha Microscopii, and Lacaille 8760. Some of the deep-sky objects are NGC 6925 and NGC 6923 Its neighboring constellations are Sagittarius, Capricornus, Grus, Indus, Pisces, and Telescopium. It is also associated with Microscopids meteor shower. The best time to view the constellation is in September.

Monoceros

Monoceros, the unicorn, is the 35th largest constellation located in the second quadrant of the Northern Hemisphere (NQ2). It occupies an area of 482sq. Degree in the sky. It was described by the Dutch cartographer Petrus Plancius in the 17th century. It represents a Unicorn, a horse-like single-horned creature.

It belongs to the Orion Family of the constellation. It is bordered by Canis Major, Canis Minor, Gemini, Hydra, Lepus, Orion, Puppis. The 3 major stars are- Alpha Monocerotis, Gamma Monocerotis, and Delta Monocerotis. Some famous deep-sky objects are Hubble's Variable Nebula (NGC 2261), Christmas Tree Cluster, Butterfly Nebula (NGC 2346), Red Rectangular Nebula (HD 44179), Cone Nebula (NGC 2264) Seagull Nebula (IC 2177), and Dreyer's Nebula (IC 447). It is associated with three meteor showers- Alpha Monocerids, December, and Monocerids. It is best visible in February.

Musca

Musca, meaning the fly in Latin, is the 77th largest in the sky. It is located in the third square quadrant of the Southern hemisphere (SQ3). It occupies an area of 138sq. Degrees in the sky.

It was created by the Dutch astronomer Petrus Plancius in the 17th century. It was first depicted by Johann Bayer in 1603 and he named it the bee. This constellation depicts an insect providing nourishment for the nearby constellation Chameleon. Chameleon is a neighboring constellation, its tongue is trying to eat the insect. Musca was named by the French astronomer Nicolas-Louis de Lacaille.

Musca belongs to the Johann Bayer family of the constellation. The 5 major stars in the constellation are Alpha Muscae, Beta Muscae, Delta Muscae, Lambda Muscae, and Gamma Muscae. Some of the notable deep-sky objects in Musca are- the Spiral Planetary Nebula (NGC 5189), Hourglass Nebula, and Dark Doodad Nebula. It is bordered by Apus, Carina, Centaurus, Chamaeleon, Circinus, and Crux. It is not associated with any meteor shower. The best time to view the constellation is in May.

Norma

Norma, meaning normal, is the 74th largest constellation in the sky. It is located in the third square quadrant in the Southern sky (SQ3). It occupies an area of 165sq. Degree in the sky. It was introduced by the French astronomer Nicolas-Louis de Lacaille in 1750. The constellation represents the carpenter's Square, a draughtsman's set square and rule. Its original name was I'Equerre et la Eagle.

Norma belongs to the Lacaille family of constellations. The major stars of the constellations are Gamma Normae, Epsilon Normae, Lota-1 Normae, Eta Normae, and Delta Normal. Some of the notable deep-sky objects are- Ant Nebula, Fine-Ring Nebula, Norma Cluster, and NGC 6152. It is bordered by Ara, Lupus, Circinus, Triangulum Australe, and Scorpius. It is associated with the Gamma Normids Meteor shower which occurs in March. The best time to view the constellation Norma is in July.

Octans

Octans, meaning the octan, is the 50th largest constellation in the sky. It is located in the fourth quadrant in the Southern Hemisphere (SQ4). It was introduced by the French astronomer Nicolas-Louis de Lacaille in the 18th century. The original name was I'Octans de Reflexion, meaning the reflecting octant.

Octans belong to the Lacaille family of Constellation. The three major stars are Nu Octantis, Beta Octantis, and Delta Octantis. The neighboring constellations are Tucana, Indus, Pavo, Apus, Chamaeleon, Mensa, Hydrus. It is not associated with any meteor shower. The best time to view the constellation is in October.

Ophiuchus

Ophiuchus, the serpent-bearer, is the 11th largest constellation in the sky. It occupies an area of 948 sq. Degree in the third quadrant of the Southern hemisphere (SQ3). It was one of the 48 constellations listed by the Greek astronomer Claudius Ptolemy in the 2nd century BC.

According to Greek Mythology, Ophiuchus is associated with Asclepius. However as reported by some sources, the Greek Mythology of Ophiuchus overlaps with the myth of Serpens.

Asclepius, the god of medicine, was the son of Apollo and Coronis. When he was in the womb of his mother, Artemis burned her for betraying Apollo. Artemis cut the open womb and rescued the baby. The boy has excellent healing skills and was regarded as the healer. He had an exceptional art of healing, surgery, and medicine. He married Epione and has three sons and six daughters. He used to hold a rod with a small wrap around it; It symbolizes the snake bite. He has a bottle of magical blood gifted by the scary Gorgons. The blood has the Magical power to kill and retrieve people. He became very famous for his miraculous cures and went too far off places to heal the parching people.

One suspicious day, a man died from excruciating pain; But Asclepius with his wicked art of healing he brought the dead man back to life. Zeus, the god of the sky, and Hades, the god of the underworld, were worried that Asclepius was the god of medicine and he knew the art of resurrection. They thought that Asclepius would teach the art of resurrection to his disciples and decided to kill him. Apollo, son of Zeus and father of Asclepius got infuriated by his father's action and concluded to avenge his father. Cyclops, who was a one-eyed creature and created Zeus' lightning bolt, was on the Apollo sights. Apollo killed Cyclops avenging his father's actions. Zeus punished Apollo by expelling him from Mount Olympus.

It belongs to the Hercules Family of the constellation.

Ophiuchus is bordered by Aquila, Hercules, Libra, Sagittarius, Scorpius, and Serpens. The major stars are- Rasalhague (Alpha Ophiuchi), Sabik (Eta Ophiuchi), Yed Prior (Delta Ophiuchi), Celbalrai (Beta Ophiuchi), Yed posterior (Epsilon Ophiuchi), and Sinistra (Nu Ophiuchi). Some of the famous deep-sky objects are- Messier 9 (NGC 6333) and Messier 10 (NGC 6254) are the globular clusters; the Little Ghost Nebula; the Dark Horse Nebula; the Pipe Nebula; the Snake Nebula; the Twin Jet Nebula; and Rho Ophiuchi Cloud Complex. It is associated with 4 meteor showers- Ophiuchids, Northern May Ophiuchids, Southern May Ophiuchids, and Theta Ophiuchids. The best time to view the constellation is in July.

Orion

Orion, the hunter, is the 26th largest constellation in the sky. It covers an area of 594sq. Degree in the first quadrant of the Northern hemisphere (NQ1). It was among the 48 constellations listed by the Greek astronomer Claudius Ptolemy in the 2nd century BC.

Orion, one of the most beautiful constellations. There are several different stories about the birth of Orion. According to one version, Orion was the son of a poor shepherd named Hyrieus. Cyrus was a generous and kind man. He treated his guest as God. Once Zeus, Hermes, and Poseidon were traveling. On the way, they visited Hyrieus' house. Hyrieus was so generous with his guest that he killed the only animal he had- an ox. Unaware of the fact that his guests were Gods; Hyrieus treated them no less than god. He served them very well. At last, Zeus revealed his true identity and decided to reward Hyrieus' generosity by granting him a wish. Hyrieus was desperate for a child. Out of his desperation, he placed his only wish in front of Zeus. The god told him to bury the hide of the bull he had sacrificed to them and pee on it. After nine months a boy was born in that place. The child became a very charming and strong man, Orion.

Orion was a hunter. Hunting was not only his work but

also the other reason for being famous. He was such a good hunter that he was hired by king Oenepion to kill the Ferocious beasts that were terrifying the inhabitants of the island Chios. Ecstatic about his success, he said he would kill all the animals on the Earth. Gaia was the goddess and mother of all animals. She was shocked by Orion's intention. To save all the animals Gaia sent an enormous Scorpion on Orion. Orion thought he would easily kill the jumbo Scorpion but soon he realized that his strength and sword were useless against the mighty beast. Realizing this, he tried to escape, but he was stung to death by the scorpion. Gaia placed Orion in the sky in front of Scorpion, so that he is being chased forever for not respecting the animals and the environment.

The brightest star of Orion is Rigel (Beta Orionis) with an apparent magnitude of 0.18 and is the 6th brightest star in the sky. The second brightest star is Betelgeuse (Alpha Orionis) with an appearance of 0.42 and is the 11th brightest star in the sky. The other major stars are- Bellatrix (Gamma Orionis), Mintaka (Delta Orionis), Alnilam (Epsilon Orionis), and Alnitak (Zeta Orionis). The notable deep-sky objects are- Orion Nebula (NGC 1976), De Mairan's Nebula (NGC 1982), Horsehead Nebula, Barnard's Loop, Flame Nebula (NGC 2024), Monkey Head Nebula (NGC 2174), Orion Trapezium Cluster, and Orion Molecular Cloud Complex.

It belongs to the Orion Family of the Constellation. The neighboring constellations are- Gemini, Eridanus, Lepus, Monoceros, and Taurus. It is associated with two meteor showers namely, Orionids and Chi Orionids. The best time to view the constellation is in January.

Pavo

The constellation Pavo, meaning peacock, is the 44th largest constellation in the sky. It is located in the fourth quadrant in the Southern hemisphere (SQ4). It covers an area of 378sq. Degree in the sky. It was introduced by the Dutch astronomer Petrus Plancius in the 16th century. It represents the Java green peacock which was Hera's sacred bird in Greek Mythology.

It belongs to the Johann Bayer Family of Constellations. The 3 major stars of the constellations are Peacock (Alpha Pavonis), Delta Pavonis, and Beta Pavonis. Apus, Chamaeleon, Dorado, Grus, Hydrus, Indus, Musca, Phoenix, Tucana, and Volans are the neighboring constellations. It contains some of the galaxies such as NGC 6744, NGC 6782, IC 4687, IC 4689, IC 4687/6, and a globular cluster- NGC 6752. It is associated with Delta Pavonids meteor showers. The constellation is best visible in August.

Pegasus

Pegasus, the winged horse, is the 7th largest constellation in the sky. It occupies an area of 1121sq. Degree. It is located in the fourth quadrant of the Northern hemisphere (NQ4). It was cataloged by the Greek astronomer Claudius Ptolemy in the 2nd century BC.

In Greek Mythology, Pegasus was an immortal winged horse. He was tamed by the hero of Bellerophon. He rode him into the battle to fight the first-breathing Khimaira. After this battle, Bellerophon attempted to ride Pegasus to heaven.

The Great Square of Pegasus is an asterism formed by the three brightest stars in the constellation of Pegasus namely Markab, Scheat, and Algenib, and the brightest star in the constellation of Andromeda is Alpheratz. This asterism is used for finding numerous deep-sky objects such as the Andromeda Galaxy, the Triangulum Galaxy, The Great Pegasus Cluster, and many more.

It belongs to the Perseus Family of the constellation. Perseus is bordered by Andromeda, Aquarius, Cygnus, Delphinus, Equuleus, Lacerta, Piscis, and Vulpecula. The major stars are- Markab (Alpha Pegasi), Scheat (Beta Pegasi), Algenib (Gamma Pegasi), Enif (Epsilon Pegasi),

Homam (zeta Pegasi), Matar (Eta Pegasi), Baham (Theta Pegasi), and Sadalbari (Mu Pegasi). Some of the famous deep-sky objects are- Cumulo de Pegaso (NGC 7078), Stephan's Quintet (HCG 92), Einstein Cross, and Propeller Galaxy. It is associated with the July Pegasids meteor shower. The best time to view the constellation is in the month of October.

Perseus

Perseus is the 24th largest constellation. It occupies an area of 615sq. Degree in the sky. It is located in the first quadrant of the Northern hemisphere (NQ1). It was in the list of 48 constellations cataloged by the Greek astronomer Claudius Ptolemy in the 2nd century BC.

Perseus was the son of Zeus and Danae. He was the one who killed Medusa, the hideous gorgon with snakes for the hair. She had been cursed. Anyone she looked at would turn into stone. After killing Medusa when Perseus was carrying her head, on his way he found Andromeda. She was to be sacrificed to the Sea monster, Cetus. Princess Andromeda was the daughter of king Cetus and queen Cassiopeia. She was the most pretty lady in the world. She was so beautiful that her mother claimed her to be more beautiful than the sea nymphs of Poseidon.

As punishment, it was decided that Andromeda must be sacrificed to the Sea monster. Perseus learned about the tragic scene and decided to save Andromeda. He then stood out to fight with the fierce sea monster. He showed his bravery and successfully killed Cetus. Then he married Andromeda and lived happily with her forever. For his bravery, he was placed in the sky in the form of a constellation.

It belongs to the Perseus Family of the constellation. The major stars are- Mirfak (Alpha Persei), Algol (Beta Persei), Atik (Zeta Persei), Delta Persei, and Theta Persei. Some of the notable deep-sky objects are- The Little Dumbbell Nebula, Alpha Persei Cluster, Perseus Molecular Cloud, Perseus Cluster, The Double Cluster, and California Nebula. The bordering constellations are- Andromeda, Aries, Auriga, Camelopardalis, Cassiopeia, Taurus, and Triangulum. It is associated with two meteor showers- Perseids and September Perseids. The best time to view the constellation is in the month of December.

Phoenix

Phoenix is the 37th largest constellation in the sky located in the first quadrant of the Southern hemisphere (SQ1). It covers an area of 469sq. Degree in the sky. It was introduced by the French astronomer Nicolas-Louis de Lacaille in the 18th century.

It represents the phoenix of mythology, which represents a bird that rises from its ashes; It was an eagle with red and gold feathers and a scarlet and gold tail. The Southern bird is the collective name for the four constellations Phoenix, Grus, Pavo, and Tucana.

Pictor

Pictor, the painter, is the 59th largest constellation in the sky located in the first square quadrant in the Southern sky (SQ1). It occupies an area of 247sq. Degree in the sky. It represents the easel and palette. It was introduced by the French astronomer Nicolas-Louis de Lacaille in the 18th century. Earlier, he named the constellation 'Equuleus Pictoris'. This constellation symbolizes the Age of Enlightenment.

Pictor belongs to the Lacaille family of constellations. The 5 major stars in the constellation are α Pictoris (Alpha Pictoris), β Pictoris (Beta Pictoris),γ Pictoris (Gamma Pictoris), δ Pictoris (Delta Pictoris) and Kapteyn's Star. Pictor contains a peculiar lenticular galaxy named NGC 1705 which is approximately 17 million light-years away from the Earth. The bordering constellations are Caelum, Columba, Carina, Dorado, Puppis, and Volans. It is not associated with any meteor shower. The best time to view the constellation is the month of January.

Pisces

Pisces, the fish, is the 14th largest constellation located in the first quadrant of the Northern hemisphere (NQ1). It occupies an area of 889sq. Degree in the sky. It was discovered by the 2nd-century Greek astronomer Claudius Ptolemy.

According to Greek Mythology, Piscis is associated with the two fishes, who were the offsprings of Piscis Austrinus. A winged monster with the head of a human and snake's body named Typhon was chasing Aphrodite, Goddess of Love, and her son Eros, God of Love. The two large fishes helped them to escape from Typhon. The two fishes were rewarded for their bravery and were placed in the sky by Goddess Athena.

It belongs to the Zodiac Family of the constellation. The major stars are- Kullat Nunu (Eta Piscium), Alrescha (Alpha Piscium), Fum al Samakah (Beta Piscium), Piscis Boreus (The North Fish), Pisces Austrinus (The South Fish). Some notable deep-sky objects in Pisces are- The Pisces Dwarf (PGC 3792), NGC 7537, and Messier 74 (NGC 628). The neighboring constellation is- Triangulum, Andromeda, Pegasus, Aquarius, Cetus, and Aries. It is associated with the Piscids meteor shower. The constellation is best visible in the month of November.

Piscis Austrinis

Piscis Austrinus, the southern fish, is the 60th largest constellation. It is located in the fourth quadrant of the Southern hemisphere (SQ4). It occupies an area of 245sq. Degrees in the sky. It was listed by the Greek astronomer Claudius Ptolemy in the 2nd century BC.

In Greek Mythology, Piscis Austrinus is known as the Great fish. It is depicted as the fish swallowing the water being poured out by the Aquarius, the water bearer. It is also considered as the parent of Piscis, the two fishes who rescued Aphrodite and Eros from Typhon.

It belongs to the Heavenly Waters Family of the constellation. The major stars are- Fomalhaut (Alpha Piscis Austrini), Beta Piscis Austrini and Gamma Piscis Austrinus. Some of the famous deep-sky objects are- NGC 7173 is an elliptical galaxy discovered by John Herschel; NGC 7174 is a spiral galaxy; and NGC 7172. The bordering constellations are Aquarius, Capricornus, Grus, Microscopium, and Sculptor. The best time to view the constellation is in the month of October.

Puppis

Puppis is the 20th constellation in the sky located in the second square quadrant in the Southern hemisphere (SQ2). It occupies an area of 673sq. Degrees in the sky. It was first cataloged by the Greek Astronomer Ptolemy when it was part of Argo Navis Constellation.

The Latin name of this constellation is the Poopdeck. It represents the stern of Argo Navis. Earlier, it used to be the part of a bigger constellation Argo Navis, which was divided into three parts in 1752 by the French astronomer Nicolas-Louis de Lacaille; the three parts were- Carina, Vela, and Puppis; in which Puppis is the largest part of the constellation.

Puppis belongs to the Heavenly Waters family of constellations. The 3 major stars of this constellation are Naos meaning ship, Tureis meaning shield, and Asmidiske meaning gunwale. Carina, Canis Major, Columba, Hydra, Monoceros, Pictor, Pyxis, and Vela are the bordering constellations. Some of the famous deep-sky objects are- Messier 46 (NGC 2437), Messier 47 (NGC 2422), Messier 93 (NGC 2447), Skull, and Crossbones Nebula (NGC 2467), and, NGC 2451I.

It is associated with three meteor showers- Pi Puppies,

Zeta Puppids, and Puppid-Velids. The best time to view the constellation is in February.

Pyxis

The constellation Pyxis, mariner's compass in Latin, is the 65th largest constellation in the sky located in the second quadrant of the Southern hemisphere (SQ2). It occupies an area of 221sq. Degree in the sky.

It was introduced by the French astronomer Nicolas-Louis de Lacaille in the 17th century. The original name of the constellation was Pyxis Nautica. It represents a nautical device once used for measuring speed and distance traveled in the sea. This constellation symbolizes the Age of Enlightenment.
It belongs to the Heavenly Waters Family of the constellation. It is bordered by Antlia, Hydra, Puppis, and Vela. The major stars in the constellations are α Pyxidis (Alpha Pyxidis), β Pyxidis (Beta Pyxidis), γ Pyxidis (Gamma Pyxidis), and T Pyxidis.

Some of the famous deep-sky objects are NGC 2818- a planetary nebula that is approximately 10,400 light-years away from our solar system; NGC 2613- a barred spiral galaxy which approximately 60 million light-years away from our Galaxy; NGC 2627- an open cluster which is 8,000 light-years distant away. It is not associated with any meteor shower. The best time to view the constellation is in the month of March.

Reticulum

Reticulum, the Latin name for small net or reticule, is a small faint constellation in the southern sky. It is the 82nd largest constellation and the 7th smallest constellation in the sky. It represents the reticule. It is located in the first quadrant of the Southern hemisphere (SQ1). It was created by the German Astronomer Isaac Habrecht II in 1621.

It is the small net at the focus of an eyepiece on the telescope which makes it possible to accurately measure the star position at night. He named it 'rhombus'. But later, a French astronomer Nicolas Louis de Lacaille named its reticulum. In 1922, it was approved by the International Astronomical Union (IAU).

It belongs to the Lacaille family of the constellation. Its neighboring constellations are Horologium, Dorado, and Hydrus. The major stars of the constellation are Alpha Reticuli, Beta Reticuli, Epsilon Reticule. The famous deep-sky object is the Topsy Turvy Galaxy (NGC 1313), It is unassociated with any meteor shower. The best time to view the constellation is in the month of January.

Sagitta

Sagitta, the arrow, is the 3rd smallest and 86th largest constellation. It occupies an area of 80sq. degree only. It lies in the fourth quadrant of the Southern hemisphere (SQ4). It was catalogued by the Greek astronomer Claudius Ptolemy in the 2nd century BC.

Sagitta is also known as "the arrow" is one of the 48 constellations identified by Ptolemy. It is the arrow used by Hercules to kill the eagle "The Aquila."
According to mythology, Prometheus moulded men and women out of clay in Gods' likeness and gave them fire that he had stolen from the Gods. This antagonized Zeus and he decided to punish Prometheus. Zeus chained him to Mount Caucasus, where the eagle perpetually mandated his liver, which would always grow again at night. Hercules found Prometheus during one of his journeys and decided to save his life. He employed his weapon "Sagitta", the arrow to kill the eagle and freed Prometheus.

Its neighbouring constellations are Aquila, Vulpecula, Hercules, and Delphinus. The three major stars are- Gamma Sagittae, Delta Sagittae, and Alpha Sagittae. The notable deep-sky object in the small constellation is the Necklace Nebula. It is not associated with any meteor shower. The best time to view the constellation is in the month of August.

Sagittarius

Sagittarius, the archer, is the 15th largest constellation in the sky. It lies in the fourth quadrant of the Southern hemisphere (SQ4). It occupies an area of 867 sq. Degree in the sky. It was cataloged by the Greek astronomer Ptolemy in the 2nd century BC.

Sagittarius is also known as Chiron, was a centaur, but he was distinct in many aspects. He was kind, humble, and generous unlike the other centaur, and taught music and medicine along with prophecy. Chiron was born half man and half horse.

Hercules and Chiron were loyal companions. On a dreadful day, he was accidentally shot with a poisoned arrow by Hercules. Despite having a vast knowledge of medicine, he failed to heal himself. Fortune didn't favor him, being immortal he could not die but endure the unbearable pain. To die he decided to help Prometheus, who was punished by God for giving fire to man. He offered Prometheus to replace him. As Chiron gave his immortality to release Prometheus, Zeus saw his solicitude and generosity and placed him in the sky as the constellation Sagittarius.

It belongs to the Zodiac Family of the constellation. The

major stars are- Kaur Australis (Epsilon Sagittarii), Nunki (Sigma Sagittarii), Kaus Media (Delta Sagittarii), Kaus Borealis (Lambda Sagittarii), Rukbat (Alpha Sagittarii), and Arkab (Beta Sagittarii). The notable deep-sky objects are- Lagoon Nebula (NGC 6523), Omega Nebula (NGC 6618), Trifid Nebula (NGC 6514), Sagittarius Cluster (NGC 6656), Sagittarius Star Cloud (NGC 6603), and Arches Cluster. The bordering constellations are- Aquila, Scutum, Serpens Caput, Ophiuchus, Scorpius, Corona Australis, Telescopium, Microscopium, Capricornus, and Indus. The constellation is best visible in the month of August.

Scorpius

Scorpius, the scorpion, is the 33rd largest constellation located in the third quadrant of the Southern hemisphere. It occupies an area of 497sq. Degree in the sky. It was listed in the list of 48 constellations cataloged by the Greek astronomer Claudius Ptolemy in the 2nd century BC.

You might have heard the story of Orion, the hunter. He was a fearless hunter. He vowed to kill all the animals on the Earth.

Gaia, the goddess of the Earth and the protector of animals, got angry by his decision and proclamation. She ordered Scorpius, the giant scorpion to kill Orion. Scorpius chased Orion and strung him to death. For his bravery, Gaia placed Scorpius in the sky in a position, chasing the Orion.

It belongs to the Zodiac Family of the constellation. The major stars are- Antares (Alpha Scorpii), Shaula (Lambda Scorpii), Acrab (Beta Scorpii), Dschubba (Delta Scorpii), Sardas (Theta Scorpii), Girtab (Kappa Scorpii), and Jabbah (Nu Scorpii). Some notable deep-sky objects are- Butterfly Cluster (NGC 6405), Ptolemy Cluster (NGC 6475), Cat's Paw Nebula (NGC 6334), Northern Jewel Box Cluster (NGC 6231), and War and Peace Nebula (NGC 6357). It is bordered by Sagittarius, Ophiuchus, Libra, Lupus, Norma, Ara, and Corona Australis. It is associated with two meteor showers- Alpha Scorpiids and Omega Scorpiids. The best time to view the constellation is the month of July.

Sculptor

The sculptor is the 36th largest constellation introduced by the French astronomer Nicolas Louis de Lacaille. It is located in the first quadrant of the Southern hemisphere (SQ1). It occupies an area of 475sq. Degree. Its Original name was Apparatus Sculptoris which means 'the sculptor's studio. It typically describes a carved head lying on a tripod table.

It belongs to the Lacaille family of a constellation. The three major stars of this constellation are Alpha Sculptoris, Beta Sculptoris, and Gamma Sculptoris. Its neighboring constellations are Aquarius, Cetus, Fornax, Grus, Phoenix, and Pisces Austrinus. Some of the famous deep-sky objects are the Sculptor Group- a group of galaxies and is approximately 12.7 million light-years away from the Milky Way Galaxy; The Sculptor Galaxy; the Cartwheel Galaxy; the Sculptor Dwarf; Pandora's Cluster; and Southern Cigar Galaxy. It is not associated with any meteor shower. The best time to view the constellation is in the month of November.

Scutum

Scutum, a shield in Latin, is the 5th smallest constellation, and the 84th largest constellation in the sky. It is located in the fourth southern Quadrant (SQ4). It was introduced by the Polish astronomer Johannes Hevelius. It occupies an area of 109sq. Degree in the sky.

Its original name was Scutum Sobiescianum, meaning the shield of Sobieski. This constellation was named by Hevelius in 1684 to commemorate the victory of King John III Sobieski on the battlefield of Vienna in 1679.

It belongs to the Hercules Constellation Family. Its neighboring constellation is Aquila, Sagittarius, Serpens Crado. The major stars are Alpha Scuti, Beta Scuti, Delta Scuti, Epsilon Scuti, and Eta Scuti. The two famous deep-sky Objects in this constellation are Wild Duck Cluster (M11, NGC 6705) and Messier 26, which is an open cluster. It is associated with the June Scutids meteor shower. The constellation Scutum is best visible in the month of August.

Serpens

Serpens is the 23rd largest constellation in the sky located in the third quadrant of the Northern hemisphere (NQ3). It occupies an area of 637sq. Degree in the sky. It was among the 48 constellations of Ptolemy.

The myth of Serpens overlaps with the myth of the constellation Ophiuchus. However, there is no genuine demonstration of the Greek mythology of the constellation.

Asclepius, the god of medicine, was the son of Apollo and Coronis. When he was in the womb of his mother, Artemis burned her for betraying Apollo. Artemis cut the open womb and rescued the baby. The boy has excellent healing skills and was regarded as the healer. He had an exceptional art of healing, surgery, and medicine. He married Epione and has three sons and six daughters. He used to hold a rod with a small wrap around it; It symbolizes the snake bite. He has a bottle of magical blood gifted by the scary Gorgons. The blood has the Magical power to kill and retrieve people. He became very famous for his miraculous cures and went too far off places to heal the parching people.

One suspicious day, a man died from excruciating pain;

But Asclepius with his wicked art of healing he brought the dead man back to life. Zeus, the god of the sky, and Hades, the god of the underworld, were worried that Asclepius was the god of medicine and he knew the art of resurrection. They thought that Asclepius would teach the art of resurrection to his disciples and decided to kill him. Apollo, son of Zeus and father of Asclepius got infuriated by his father's action and concluded to avenge his father. Cyclops, who was a one-eyed creature and created Zeus' lightning bolt, was on the Apollo sights. Apollo killed Cyclops avenging his father's actions. Zeus punished Apollo by expelling him from Mount Olympus.

It belongs to the Hercules Family of the constellation. The major stars are – Unukalhai (Alpha Serpentis), Alya (Theta Serpentis), Eta Serpentis, and Beta Serpantis. Some major deep-sky objects are- The Eagle Nebula (NGC 6611), Seyfert's sextet (NGC 6027), Red Square Nebula, and Hoag's Object. The neighboring constellations are- Corona Borealis, Boötes, Virgo, Libra, Hercules, Aquila, Ophiuchus, Sagittarius, and Scutum. It is not associated with any meteor shower. It is best visible in the month of July.

Sextans

Sextans, which means astronomical sextant, is the 47th largest constellation sky as it covers an area of 314 sq. Degree in the sky. It lies in the second quadrant of the Southern hemisphere (SQ2). It was introduced by the Polish astronomer Johannes Hevelius in 1687.

The original name of the constellation is Sextans Uraniae. It is named after an instrument that was used to measure the position of stars in the night. It was known to both Greeks and Ptolemy.

It belongs to the Hercules Family of the constellation. The main stars in the constellation are Alpha Sextans: Gamma Sextantis and Beta Sextantis. Some of the notable deep-sky objects are- The Spindle Galaxy (NGC 3115); Sextans Dwarf Spheroidal; Cosmos Redshift 7. It is bordered by Leo, Hydra, and Crater. It is located near the Celestial Equatorial and can be seen from both hemispheres. It is associated with the Sextantids meteor shower. It is best visible in the month of April.

Taurus

Taurus, which means the "bull" In Latin, is the 17th largest constellation. It lies in the first quadrant of the Northern hemisphere (NQ1) and occupies an area of 797sq. degree. It was depicted by the Greek astronomer Ptolemy in the 2nd century BC.

Zeus took the incarnation of the bull to seduce Europa, the daughter of Agenor. When he stood in front of Europa, he was the most abundant bull. Europa decided to sit on his back. When she sat, he flew away with the lovely princess. When they reached Crete, Zeus transformed himself and revealed his genuine identity. They lived happily and ended up having three sons. Among his sons, Minos was the eminent one. He became the king of Crete and built a Knossos at his place, where bull games were held every year. Out of eternal happiness, Zeus placed the constellation of the bull high in the boundless sky.

It belongs to the Zodiac Family of the constellation. The bordering constellation is Aries, Auriga, Cetus, Eridanus, Gemini, Orion, and Perseus. The major stars of the constellation are- Aldebaran (Alpha Tauri) is the brightest star of the Taurus with an apparent magnitude of 0.85 and is the 13th brightest star; Elnath (Beta Tauri); Pectus Tauri (Lambda Tauri); Ain (Epsilon Tauri); Hyadum I (Gamma

Tauri); Alcyone (Eta Tauri). The notable deep-sky objects in the constellation are- The Crab Nebula (NGC 1952), The Pleiades (M45); The Hyades Cluster (Caldwell 41); The Hind's Variable Nebula (NGC 1555); Crystal Nebula (NGC 1514); The Merope Nebula (NGC 1435). It is associated with two meteor showers- Taurids and Beta Taurids. It is best visible in the month of January.

Telescopium

Telescopium, the Telescope, is the 57th largest constellation in the sky. It is the fourth quadrant in the Southern hemisphere (SQ4). It is among the 12 constellations named by French astronomer Nicolas-Louis de Lacaille in the 18th century.

The constellation was first discovered in 1752 by Lacaille from the southernmost tip of Africa, Cape Of Good Hope. It is unassociated with any myths. It represents a Refractor telescope; it is also known as Tubus Atronomicus.

It belongs to the Lacaille family of the constellation. Its neighboring constellations are Ara, Corona Australis, Indus, Microscopium, Pavo, Sagittarius. The 5 major stars of the constellation are Alpha Telescopii, Zeta Telescopii, Epsilon Telescopii, Lambda Telescopii and Lota Telescpii. One of the notable deep-sky objects in the constellation is the Telescopium Group (AS0851) a group of 12 galaxies that are approximately 120 million light-years away from the Milky Way Galaxy. It is not associated with any meteor shower. The best time to view the constellation is in the month of August.

Triangulum

Triangulum, Triangle in English, is the 78th largest constellation that lies in the first quadrant of the Northern Hemisphere. It occupies an area of 132sq. Degree in the sky. It was first cataloged by the 2nd-century Greek astronomer Ptolemy in his list of 48 constellations. It was entitled 'Tripertitas' by Johann Bayer in the 17th-century.

It belongs to the Perseus Family of the constellation. The major stars are- Caput Trianguli (Alpha Trianguli), Beta Trianguli, Gamma Trianguli, and Delta Trianguli. The notable deep sky objects are- The Triangulum Galaxy (NGC 598)- a spiral galaxy with an apparent magnitude of 5.72; NGC 634- a spiral; NGC 604- an emission nebula discovered by William Herschel; NGC 784- a barred spiral galaxy with an apparent magnitude of 12.23; and NGC 953- an elliptical galaxy discovered by Heinrich Louis d'Arrest and has an apparent magnitude of 14.5.

It is bordered by Andromeda, Pisces, Aries, and Perseus. It is not associated with any meteor showers. The best time to view the constellation is the month of December.

Triangulum Australe

Triangulum Australe, the southern triangle, is the 83rd largest constellation in the sky. It covers an area of 110sq. Degree and lies in the third quadrant of the Southern Hemisphere. It represents surveying instruments. It was created by the Dutch cartographer Petrus Plancius in the 16th century and was first depicted in the Uranometria, Johann Bayer's star atlas. It was also called "le Triangle Austral ou le Niveau" by Nicolas Louis de Lacaille in 1756.

It belongs to the Hercules Family of the constellations. The major stars of the constellation are- Atria (Alpha Trianguli Australis), HD 133683, HD 147018. It contains a planetary nebula called Henize 2-138, and a pair of galaxies called ESO 69-6. The neighboring constellations are Norma, Ara, Circinus, and Apus. It is not associated with any meteor shower. It is best visible in the month of July.

Tucana

Tucana is the 48th largest constellation in the sky. It is named after a bird called toucan. It was among the 12 constellations introduced by Petrus Plancius. It is located in the first quadrant of the Southern hemisphere (SQ1) and covers an area of 295sq. Degree in the sky.

The constellation is named after a colourful huge bill, the South American bird toucan. It was published in the star atlas of Johann Bayer named Uranometria in the 16th century. Its original name was Den Indiaenschen Exster, op indies Lang ghenaemt and was shortened to Tucana.

The major stars in the constellation are: α Tucanae (Alpha Tucanae) is the brightest star with an apparent magnitude of 2.86m. γ Tucanae (Gamma Tucanae) is the second brightest star; it has an apparent magnitude of 3.99m. ζ Tucanae (Zeta Tucanae) is the third brightest star with an apparent magnitude of 4.23m. Some of the other stars are ϰ Tucanae (Kappa Tucanae), β Tucanae (Beta Tucanae), ε Tucanae (Epsilon Tucanae), δ Tucanae (Delta Tucanae), ν Tucanae (Nu Tucanae), ι Tucanae (Iota Tucanae). Some of the deep-sky objects are- the Tucana Dwarf, the Small Magellanic Cloud, two open clusters NGC 290 and NGC 265, and a globular cluster NGC 104 – 47 Tucanae.

Tucana belongs to the Johann Bayer family of constellations. It is bordered by Grus, Indus, Octans, Hydrus, Eridanus, Phoenix. This constellation is not associated with any meteor shower. The best time to view the constellation is in the month of November.

Ursa Major & Ursa Minor

Ursa Major, the great bear, is the 3rd largest constellation in the sky. It covers an area of 1280sq. Degree in the and is located in the second quadrant of the northern hemisphere.

Ursa Minor, the lesser bear, is the 56th largest constellation located in the third quadrant of the northern hemisphere. It covers an area of 256sq. Degree. Both Ursa Major and Ursa Minor were cataloged by the Greek astronomer Claudius Ptolemy in the 2nd century BC.

Callisto, a beautiful princess, and daughter of King of Arcadia was best known for her elegance. She attracted both men and Gods. The elegant princess became the lover of Zeus, and a boy Arcus was born from their union.

But miseries and distress were not far. After learning of another betrayal of Zeus, her wife Goddess Hera decided to punish Callisto. Hera cursed the princess and then fur started to appear on the young princess's soft skin. Her delicate hand turned into paws and sharp claws. The princess tried to screen for help but only a dreadful howl came out of her mouth. Callisto had been transformed into the great bear. The bear wandered through the forest and stood on her two paws and begged Zeus to have her original shape back. Zeus did not want to antagonize his wife and did nothing for Callisto.

The bear was now roaming around the bank of the forest, next to where she lived. She used to hide from everyone and watched her friends and relatives. Despite the fear of being snipped by the hunter, she only risked her life to see her son Arcus grow. The little boy grew up believing that one day he would see his mother again. Time passed and one day while wandering through the forest, she encountered a hunter who was her beloved son. The mother forgot that she was a bear and stood up and tried to hug her son. Arcus thought that the bear was about to attack him and thrust the spear into the bear. But Zeus did not allow such a tragedy to occur. God transformed both mother and son into a constellation. The mother became known by the name of Ursa Major, and her son's constellation was Ursa Minor.

But the Goddess Hera was furious to see her rival and her son being honored in the skies. She went and complained to Poseidon. Therefore, the constellations Ursa Major and Ursa Minor cross heaven, throughout the night without hiding into the seawater for a single moment. Even then Callisto's bliss would not diminish. Thus, she would spend eternity with her son.

Ursa Major belongs to the Ursa Major Family of the constellation. The brightest star of the constellation is called Alioth. Its bordering constellations are Draco, Camelopardalis, Lynx, Leo, Leo Minor, Coma Berenice, Canes Venatici, and Boötes It is associated with two meteor showers- Alpha Ursa Majorids and Leonids-Ursids. It is best visible in April.

Ursa Minor belongs to the Ursa Major family of the constellation. It contains the most significant star named Polaris. Polaris is the brightest star of the constellation. The big dipper and Ursa Major are about Polaris. It is also called the Pole star. Its bordering constellations are Draco, Camelopardalis, and Cepheus. It is associated with the Ursids meteor shower. The constellation is best visible in June.

Vela

Vela, the Latin name is the sail of a ship, is the 32nd largest constellation in the southern sky. It occupies an area of 500sq. Degree in the sky. It is located in the second quadrant of the Southern hemisphere (SQ2).

The sail of vela can be seen with the naked eye as it has an apparent magnitude of 1.8m. Earlier, it used to be part of a Greater constellation, Argo Navis. Later in 1752, Argo Navis was divided into three smaller constellations by the French Astronomer Nicolas Louis de Lacaille; the other constellations were Carina and Puppis.

Vela belongs to the Heavenly Waters Family Of Constellation. Some of the major stars of the constellation are Regor (Gamma Velorum, Alsephina (Delta Velorum), Suhail (Lambda Velorum), and Markeb (Kappa Velorum). Some of the famous deep-sky objects are- The Eight-Burst Nebula (NGC 3132), the Pencil Nebula (NGC 2736), Gum Nebula, Omicron Velorum Cluster, Vela Supernova Remnant, and Vela Pulsar. It has three meteor showers, namely, Delta Velids, Gamma Velids, Puppid-Velids. The bordering constellations are Antlia, Pyxis, Puppies, Carina, Centaurus. The best time to view the constellation is in the month of March.

Virgo

Virgo, the virgin, is the 2nd largest constellation occupying an area of 1294sq. Degree in the sky. It lies in the third quadrant of the Southern hemisphere (SQ3). It was listed in the list of the 48 constellations cataloged by the Greek astronomer Claudius Ptolemy in the 2nd century BC.

According to Greek Mythology, The Virgo constellation is associated with Goddess Astraea, the virgin goddess of Justice, Innocence, and Purity. She was the beloved daughter of Zeus and Themis. Her remarkable beauty was blessed by Aphrodite, the goddess of Love, and Her power was blessed by Hermes. Zeus sent her down to the earth during the mythical Golden age to punish the men.

Epimetheus was astonished by her beauty and gave her a box named Pandora as a gift from God. Astraea cannot resist herself to open the box. When she unlocks the box all the evil powers came out of it and were delivered to the men.

After this event, Zeus placed her in the sky among the stars.

It belongs to the Zodiac Family of the Constellation. Its neighboring constellations are Boötes, Coma Berenice,

Leo, Crater, Corvus, Hydra, Libra, and Serpens Caput. The major stars are- Spica (Alpha Virginis) is the brightest star in Virgo with an apparent magnitude of 1.04, it is 15th brightest star in the sky; Porrima (Gamma Virginis), Auva (Delta Virginis), Vindemiatrix (Epsilon Virginis), Heze (Zeta Virginis), and Zaniah (Eta Virginis).

Some notable deep-sky objects located in the constellation are- Virgo Cluster, Sombrero Galaxy, Eyes Galaxies, and Siamese Twins. It is associated with two meteor showers- Virginids and Mu Virginids. It is best visible in the month of May.

Volans

Volans is one of the smallest constellations in the southern sky. It occupies an area of 141sq. Degrees in the sky. It is among the 12 constellations introduced by the Dutch navigator Petrus Plancius in the 16th century. It is in the second quadrant in the Southern hemisphere (SQ2). The original name of the constellation is Pisces Volans, the flying fish. It is the 76th largest constellation in the sky.

It was introduced by the Dutch navigators Pieter Dirkszoon Keyser and Frederick de Houtman in the late 16th century. And it was depicted on Petrus Plancius' globe in 1598. Plancius came as constellation Vliegendenvis, meaning Flying Fish. It can jump out of the water and glide through the air on wings.

It belongs to the Johann Bayer family of the constellation. The major stars are- Beta Volantis, Gamma Vola. Some of its brightest stars are HR 3347, HR2736, HR3024, HR2803, HR3615, HR3223, HR3280, HR2662, HR3460, HR3334, HR3537, HR3301. Its neighboring constellation is Carina, Pictor, Dorado, Chamaeleon, and Mensa. It is not associated with any meteor shower. It is best visible in the month of March.

Vulpecula

Vulpecula, the Latin name is a little fox, was introduced by Polish astronomer Johannes Hevelius in 1687. It lies in the fourth quadrant of the Northern hemisphere (NQ4). It is located in the middle of the summer triangle. It is the 55th largest constellation in the sky as it occupies an area of 268sq. Degrees.

This constellation does not have a myth constellation. According to Hevelius, the constellation depicted the double figure of a fox, Vulpecula, and a goose, Anser. The fox is holding the goose in his mouth. The goose has been eaten by the fox. Hevelius says the fox was taking the goose to his neighboring Cerberus. Cerberus was the constellation invented by Hevelius, though it is now obsolete.

It belongs to the Hercules Family of the constellation. Its neighboring constellations are Cygnus, Lyra, Hercules, Sagitta, Delphinus, and Pegasus. The 3 major stars are Alpha Vulpeculae, 23 Vulpeculae, 31 Vulpeculae. Some notable deep-sky objects are- The Dumbbell Nebula and the Brochhi's Cluster. It is best visible in the month of September.

www.ingramcontent.com/pod-product-compliance
Lightning Source LLC
La Vergne TN
LVHW050412160726
843469LV00041B/1045

9789354585166